Electron Paramagnetic Resonance

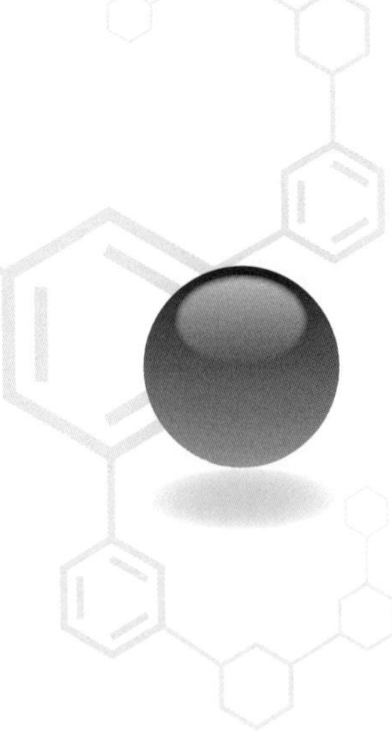

Electron Paramagnetic Resonance

Victor Chechik

Emma Carter

Damien Murphy

Great Clarendon Street, Oxford, OX2 6DP,
United Kingdom

Oxford University Press is a department of the University of Oxford. It furthers the University's objective of excellence in research, scholarship, and education by publishing worldwide. Oxford is a registered trade mark of Oxford University Press in the UK and in certain other countries

Impression: 1

Published in the United States of America by Oxford University Press
198 Madison Avenue, New York, NY 10016, United States of America

British Library Cataloguing in Publication Data
Data available

Library of Congress Control Number: 2015960996

ISBN 978-0-19-872760-6

Printed in Great Britain by Bell & Bain Ltd., Glasgow

Preface

Electron Paramagnetic Resonance spectroscopy is a powerful and versatile technique for the characterization of paramagnetic species in many fields such as biology, chemistry, materials science, and physics. A thorough analysis of the EPR spectrum can provide an unparalleled and comprehensive description of the chemical identity, structure, symmetry, electronic properties, and dynamics of the spin system. With the advent of smaller, bench-top instruments now widely available, EPR is becoming more routinely used in many research laboratories. Whilst many spectroscopic and physical chemistry textbooks may provide a short section on EPR, and with several excellent advanced textbooks on the subject, there are few introductory texts available for a novice user of the technique. This book is intended to be an introductory level text, explaining the basic principles and use of continuous wave EPR, with a strong emphasis on the interpretation of spectral parameters and how these can be linked to chemical and physical properties of the paramagnetic species. A brief introduction to pulsed EPR methods will also be presented.

The book is primarily aimed at advanced undergraduate students. However, we hope it will also serve as a useful entry level text for postgraduate students and other novice users of EPR spectroscopy. Very often, EPR is used in the context of collaborations between EPR spectroscopists and other scientists. We envisage that our book will be useful for the non-spectroscopists involved in such collaborations, where the expert knowledge of the technique is not required, but basic understanding of the underlying principles and the scope of applications is highly beneficial.

It is unavoidable that some elements of the theory described in this book rely on prior knowledge of symmetry and group theory, operators and matrix algebra, basic organic and inorganic chemistry. However, we hope that the main concepts of the technique can be understood without advanced knowledge of these areas and, where appropriate, the reader is referred to undergraduate textbooks, including other *Oxford Chemistry Primers*. In particular, many aspects of EPR theory are very similar to NMR, and the understanding of EPR theory can be enhanced by consulting the two NMR textbooks in the *Oxford Chemistry Primers* series.

Finally, we would also like to acknowledge the expert input from numerous colleagues including G. Jeschke (ETH, Zurich), S. Van Doorslaer (University of Antwerp), Z. Sojka (Jagiellonian University, Krakow), C. Wedge (University of Warwick), and P. Knowles, C. Morley, D. Willock and A. Folli (Cardiff).

Cardiff and York
November 2015
V.C., E.C., D.M.M.

Contents

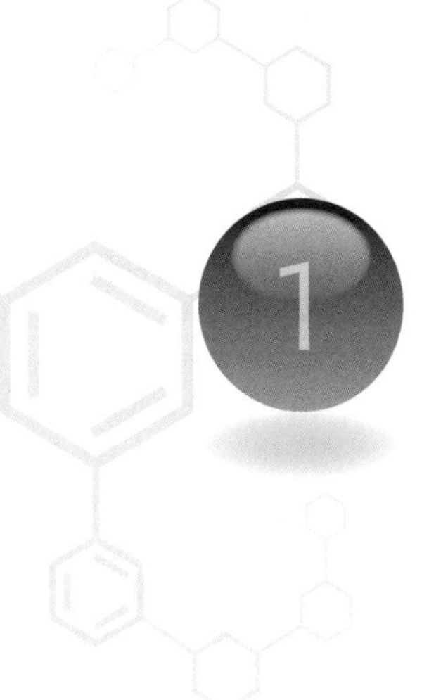

1 A brief overview of Electron Paramagnetic Resonance spectroscopy

1.1 Introduction

Electron Paramagnetic Resonance spectroscopy (abbreviated to EPR) is a magnetic resonance technique used for the study of systems containing unpaired electrons. Such systems are paramagnetic and attracted by magnetic fields. One might immediately assume that only a limited number of systems can therefore be studied by EPR spectroscopy. Fortunately, this is not the case and systems with unpaired electrons are abundant. A few examples of such systems, some of which are shown in Fig. 1.1, include:

Organic radicals in solution These are often very unstable and EPR is an important tool for detecting them. Some organic radicals are quite stable (e.g. TEMPO, see Chapter 4) and may be used as spin probes.

Organic radicals in solids These can also be quite stable, as the rigid environment prevents their diffusion and hence their termination. For instance, grinding sugar with a pestle and mortar can break some of the bonds within the molecule, generating free radicals which persist for a long time. Burnt toast is another example of a system containing an organic radical trapped inside a solid matrix and hence relatively stable. In fact, many coals and chars (including carbon nanotubes) have an inherent background EPR signal due to trapped unpaired electrons.

Transition metal ions These often contain unpaired electrons and can be studied by EPR both in solution and in the solid state. EPR spectra can provide essential structural information about the ions and their complexes with organic ligands (see Chapter 6). The gemstone ruby possesses a vibrant red colouration due to paramagnetic chromium ions which are observable by EPR.

Inorganic free radicals These are very common in nature. For instance, exposure of tooth enamel to ionizing radiation produces $CO_2^{\bullet-}$ radicals which can persist in the surrounding matrix (hydroxyapatite) for a very long time. This persistence is actually exploited in EPR dosimetry applications.

Atomic and molecular gases These may possess unpaired electrons (such as the ground state of molecular oxygen, which exists as a triplet, see Chapter 7) and hence can be studied by EPR. For example, hydrogen atoms can be detected by EPR in a burning mixture of gaseous hydrogen and oxygen.

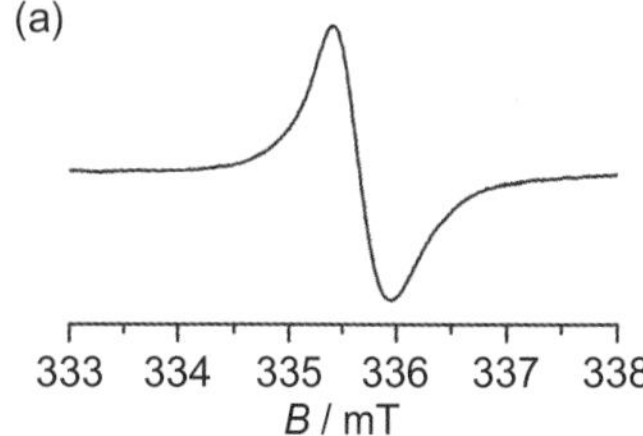

Fig. 1.1(a) EPR spectrum of brown human hair.

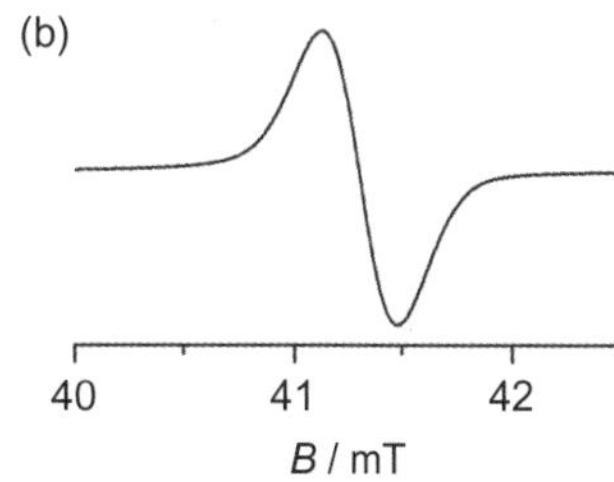

Fig. 1.1(b) EPR spectrum of an irradiated tooth recorded at 1.2 GHz.

After B. B. Williams, R. Dong, M. Kmiec, G. Burke, E. Demidenko, D. Gladstone, R. J. Nicolalde, A. Sucheta, P. Lesniewski, and H. M. Swartz, *Health Phys.* 2010, **98**, 327.

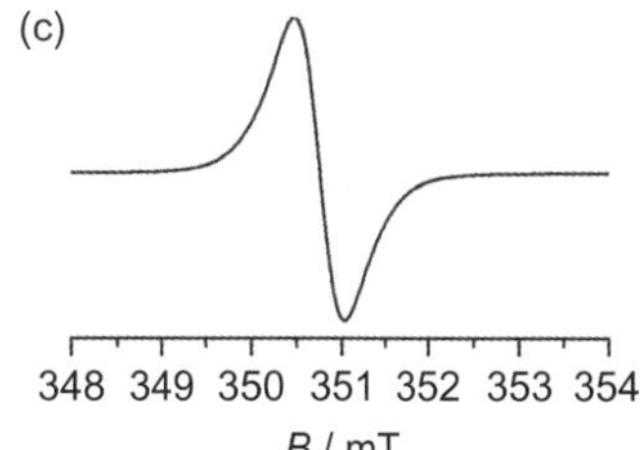

Fig. 1.1(c) EPR spectrum of heavily burnt toast.

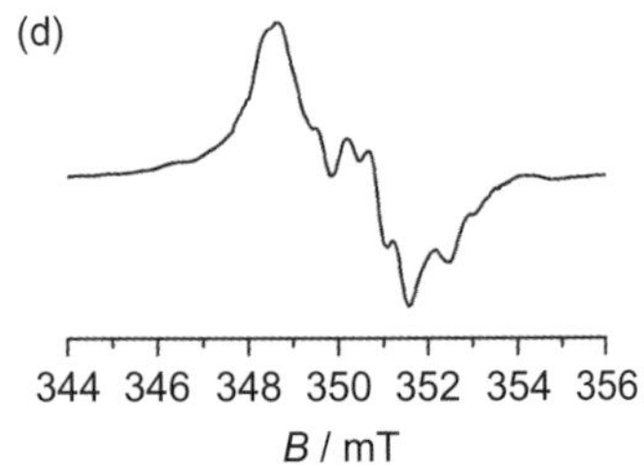

Fig. 1.1(d) EPR spectrum of sugar powder obtained by grinding granulated sugar.

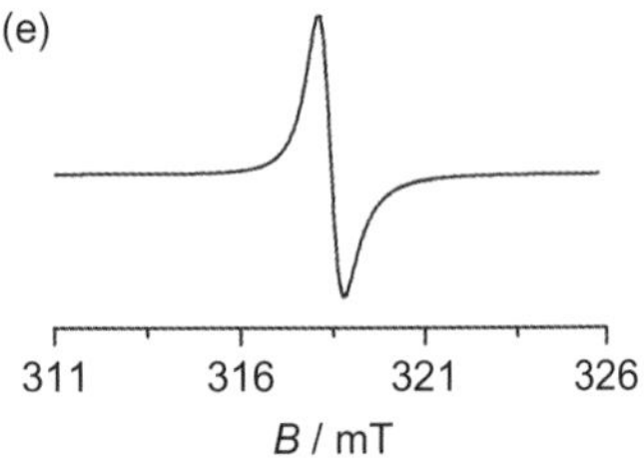

Fig. 1.1(e) EPR spectrum of a natural brown diamond.

After J.-R. Kim, Y.-D. Jang, S.-H. Kim, J.-H. Kim, and Y.-K. Paik, *J. Mineral. Soc. Korea* 2008, **21**, 435.

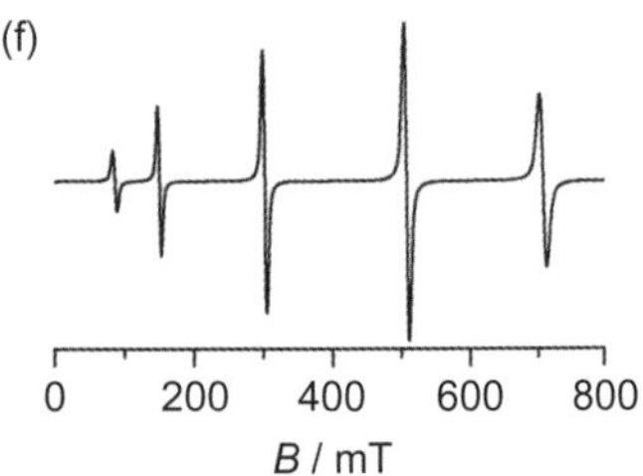

Fig. 1.1(f) EPR spectrum of Cr(III) and Fe(III) centres in natural ruby.

Unless otherwise stated, all of the EPR spectra shown in this book were recorded or simulated at X-band frequency (9.5 GHz); see Chapter 3, Table 3.1.

Wavelength and frequency are related by $\lambda\nu = c$.

The energy of a single photon is given by $E = h\nu$, where h = Planck constant = $6.626\,070 \times 10^{-34}$ J s and $\hbar = h/2\pi = 1.054\,571 \times 10^{-34}$ J s.

Biological samples Many biological systems also contain paramagnetic compounds, either as intermediates or as cofactors in enzymes (e.g. Cu(II), Mn(II), Fe(III)). EPR is an essential tool for understanding the structural features of such systems. Another example is human hair which produces a strong EPR signal attributed to melanin (a pigment that gives hair its colour, and is also responsible for the colour of skin and eyes).

Photoexcited molecules Photoexcitation of diamagnetic molecules results in the formation of an excited singlet state which is often followed by intersystem crossing to the paramagnetic triplet state that can be studied by EPR (as well as consequent radical reactions). Photovoltaic materials are another important example of materials that fall into this category for EPR detection.

Point defects in solids Ion vacancies or interstitial ions are often paramagnetic and EPR is one of the most powerful tools for their structural characterization. For instance, nitrogen is the most common impurity in naturally-occurring diamonds and the colour is believed to be related to different types of nitrogen defects.

Therefore, although the applications of EPR are confined to systems bearing unpaired electrons, it is a powerful technique that can be applied to a wide variety of gaseous, liquid, and solid samples.

1.2 The EPR technique

The basic principles and underlying physics of EPR are very similar to those encountered in NMR. Both techniques deal with the interaction of electromagnetic radiation with inherent magnetic moments within the sample. Whilst virtually all modern NMR measurements are performed in pulsed mode, both continuous wave (CW) and pulsed mode operations are common in EPR. This book will deal with CW EPR, in which low power microwaves of fixed frequency (the most common frequency is 9.5 GHz, called X-band) continuously irradiate a sample within a resonator while the applied magnetic field is swept (see Chapter 2). Pulsed EPR methods are introduced in Chapter 9.

All spectroscopic techniques are concerned with the interaction of electromagnetic radiation with matter. Excitation between two states or energy levels is induced by absorption of the radiation. The electromagnetic (EM) radiation is composed of oscillating electric ($\boldsymbol{E}_1$) and magnetic ($\boldsymbol{B}_1$) fields that propagate through space with a constant speed (c) (Fig. 1.2). In plane polarized radiation, the $\boldsymbol{E}_1$ and $\boldsymbol{B}_1$ fields are perpendicular to each other and vary sinusoidally forming a transverse wave with a specific wavelength (λ) and frequency (ν). The radiation is composed of photons which carry energy and angular momentum which can interact with matter. When the energy of the photon matches the energy separation between the two states, resonance absorption occurs.

In most spectroscopic techniques (e.g. in IR spectroscopy), the $\boldsymbol{E}_1$ field interacts with permanent or fluctuating electric dipole moments in the sample. EPR and NMR are somewhat unique in this respect, since in these techniques the $\boldsymbol{B}_1$ field usually interacts with permanent magnetic dipole moments in the sample. In NMR, the permanent magnetic dipoles arise from the nuclei, whereas it is the unpaired electrons that are responsible for the permanent magnetic dipole moments in EPR. An external

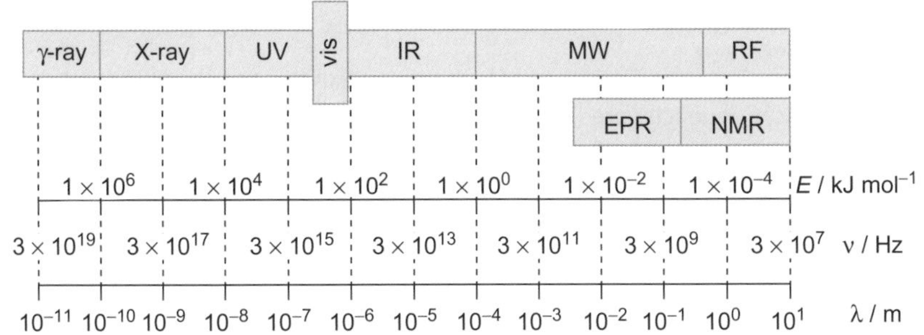

Fig. 1.3 Regions of the electromagnetic spectrum and associated energies, frequencies, and wavelengths.

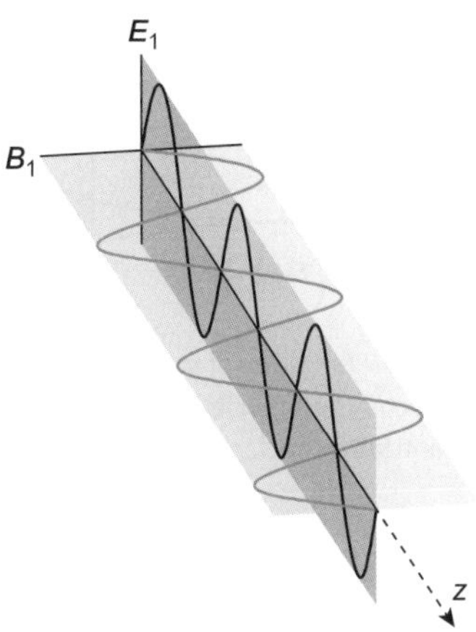

Fig. 1.2 Plane polarized EM waves showing the perpendicular electric $\boldsymbol{E}_1$ and magnetic $\boldsymbol{B}_1$ fields propagated along the *z* axis.

laboratory magnetic field (***B***) is usually applied to the system to align these dipoles, so when EM radiation of the correct frequency is available, resonance absorption will occur. In EPR, the sample is typically exposed to the $\boldsymbol{B}_1$ field of the EM radiation of fixed frequency, and the external magnetic field (***B***) is varied or swept. This is the basic premise of the EPR experiment, arising from the Zeeman splitting of the two electron spin states (see Chapter 2).

Just like NMR, EPR transitions are characterized by very small energy gaps (and consequently very long wavelengths) compared to other types of spectroscopies (Fig. 1.3). This has important consequences for the technique. For instance, the low-energy radiation used in EPR does not damage the sample (e.g. unlike X-ray techniques) so it is a non-invasive spectroscopic method. Furthermore, the populations of the different energy levels probed by EPR and NMR are very similar, as the energy gap between them is small. Hence the sensitivity of both EPR and NMR techniques is limited.

The Bohr magneton (μ_B) is the unit expressing the magnetic moment of the electron, whereas the nuclear magneton (μ_N) expresses the magnetic dipoles of heavy particles such as nuclei. Since μ_B is much larger than μ_N, this means that the magnetic moment is also larger. As a consequence of this, EPR is more sensitive than NMR, whilst resonance absorption occurs at higher frequencies and the timescales involved in EPR are also much shorter.

$$\mu_B = \frac{e\hbar}{2m_e} = 9.274 \times 10^{-24}\ \mathrm{J\ T^{-1}};$$

$$\mu_N = \frac{e\hbar}{2m_p} = 5.051 \times 10^{-27}\ \mathrm{J\ T^{-1}}$$

The following conventions are used throughout the book:

Vector parameters are bold and italicized (e.g. ***B***).

Matrices and tensors are in straight bold font (e.g. **A**).

All other parameters are italicized, with the exception of Greek letters.

Operators have hats (e.g. $\hat{H}$), vector operators are bold with hats (e.g. $\hat{\boldsymbol{S}}$).

EPR versus ESR

The fundamental origin of the magnetic moment comes from the spin angular momentum of the unpaired electron (see Chapter 2). Resonance occurs when the sample bearing unpaired electrons absorbs sufficient energy to induce a spin transition from one energy level to the next. In this situation the spectrum can be regarded as being dominated by spin angular momentum, as commonly observed in organic radicals (see Chapter 4). For such systems, the acronym ESR (Electron Spin Resonance) spectroscopy is frequently used. However, residual orbital angular momentum can also contribute to the net magnetic moment of the electron (see Chapter 5). For these systems, particularly for d- and p-block elements (see Chapter 6) and multiple-spin systems (see Chapter 7), the more complete acronym of EPR spectroscopy is used. Often, however, the acronyms ESR, EPR, and EMR are used interchangeably in the literature. Throughout this book, the acronym EPR will be used.

The acronym EMR (Electron Magnetic Resonance spectroscopy) is also occasionally used, to include not just ESR/EPR, but also the wider family of advanced techniques such as ENDOR, ESEEM, HYSCORE, DEER (see Chapter 9).

1.3 The origins

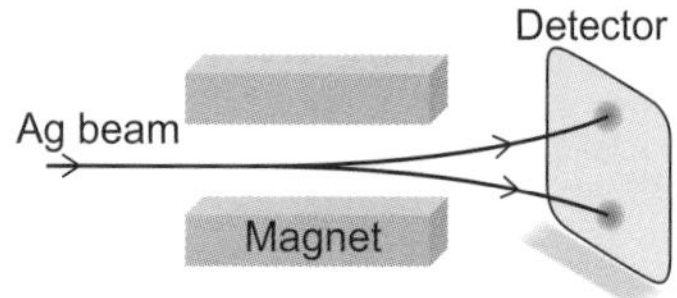

Fig. 1.4 Illustration of the Stern–Gerlach experiment evidencing the quantization of electron spin.

The high resolution atomic spectrum of hydrogen reveals that the spectral lines do not have the exact frequencies predicted by the Schrödinger wave equation. An explanation for this was proposed by the Dutch-American physicists Goudsmit and Uhlenbeck based on the spin of the electron. The electron has two spin states, denoted 'spin-up' (↑) or 'spin-down' (↓), which are characterized by the fourth quantum number (m_S), called the *spin magnetic quantum number*. The existence of the two spin states was demonstrated in an experiment performed by Stern and Gerlach in 1922. When a beam of Ag atoms was passed through a magnetic field, the beam was found to split into two (Fig. 1.4). This was attributed to the odd number of electrons in the Ag atoms, which behave like the hydrogen atom with one unpaired electron. The unpaired spin acts like a tiny magnet which is deflected through the laboratory magnet. The initial single beam splits into two beams because the 'spin-up' electrons are aligned in one direction, while the 'spin-down' electrons are aligned in another direction.

Using a modified version of the Stern–Gerlach experiment, Rabi later showed in the 1930s that by varying the magnetic fields, transitions between spin states could be induced using radio frequency (RF) fields. The first experimental observation of an EPR resonance was then performed by Zavoisky in 1945 using $CuCl_2.2H_2O$ at 4.76 mT and 133 MHz.

1.4 The EPR spectrum

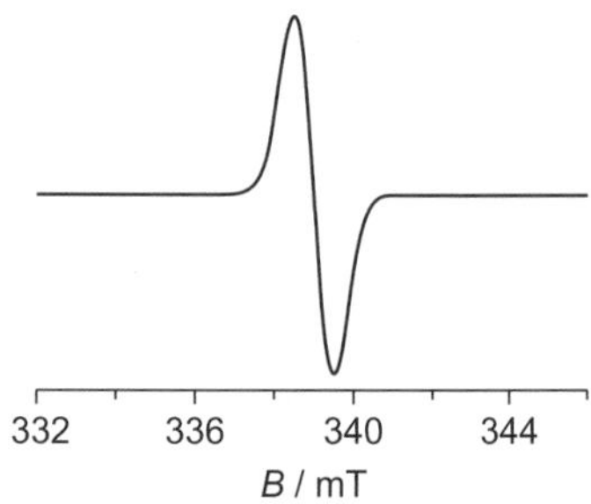

Fig. 1.5 A simple CW EPR signal recorded at 9.5 GHz for a radical with a *g* value of 2.0023.

An EPR spectrum is a plot of microwave absorption as a function of applied magnetic field intensity (*B*) (Fig. 1.5). In the CW experiment, the resonance absorption signal is plotted as a first derivative rather than in absorption mode (see Chapter 3 for details).

The EPR spectra of real systems are considerably more complex compared to the relatively simple spectra shown in Figs. 1.1 or 1.5. For example, the isotropic liquid phase spectra of organic radicals (see Chapter 4), the anisotropic frozen solution spectra of transition metal ions (see Chapter 6), or the high-spin spectra of multiple unpaired electron systems (see Chapter 7) can all at first sight appear incredibly convoluted and varied. Regardless of the complexities, one must look for and recognize certain key features in order to interpret the spectra of simple $S = ½$ spin systems (i.e. paramagnetic systems with a single unpaired electron). For CW EPR, some of the more important features are: i) the position; ii) the separation; iii) the number of lines; iv) the linewidth; and v) the intensity (Fig. 1.6).

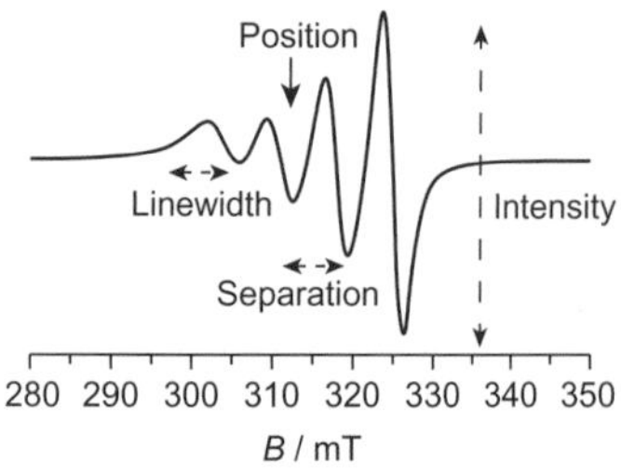

Fig. 1.6 Key distinguishing features to recognize and identify in a CW EPR spectrum.

The position is determined by the *g value* (see Chapter 2), the distance between the lines is determined by the *hyperfine* or *a value* (see Chapter 2), while the number of lines in the *hyperfine pattern* will be dictated by the nuclear spin *I*. In some cases a *fine* pattern may be the cause of the multiple lines (see Chapter 7). Numerous factors may contribute to the linewidths, including dynamic effects (see Chapter 8) or unresolved superhyperfine (see Chapter 6) or hyperfine (see Chapter 8) interactions. Finally, the differing line intensities may also depend on various influences, including the multiplicity of the hyperfine lines caused by multiple interactions with nuclei of $I \geq ½$ (see Chapter 4) or the anisotropy of the paramagnetic system (see Chapter 5).

The complexity of EPR spectra from real samples hides the wealth of information that can be obtained about the system. The EPR practitioner must therefore not only obtain the 'best quality' (i.e. resolved) spectrum, by optimizing the measurement conditions, including both sample and spectrometer (see Chapter 3), but also analyse the spectrum to provide a robust interpretation of the observed parameters. In this book, we will explain how this is achieved for common systems studied by CW EPR.

1.5 The scope and applications of EPR

Although the number of systems containing unpaired electrons that are amenable to study by EPR may be relatively small compared to the multitude of diamagnetic systems studied by NMR, this limitation can also be advantageous since the signals can be studied without any background interference. The versatility of the technique (at conventional frequencies) is enhanced by the ability to measure signals in liquid, frozen solution, and single crystal forms. Analysis of the spectra acquired from these phases provides insights into the range of magnetic interactions accessible by EPR, which in turn reveal details about the electronic properties of the system.

These interactions include: i) the electron Zeeman interactions (abbreviated EZ), caused by the electron spin interacting with the applied field; ii) the zero-field splitting (ZFS) caused by strong electron spin–electron spin interactions; iii) the hyperfine interactions (HF) caused by the electron spin interacting with the nuclear spins; iv) the nuclear Zeeman (NZ) and nuclear quadrupole (NQ) interactions caused by the nuclear spin interacting with the applied field and the nuclear spin interacting with electric field gradients in the nucleus, respectively; and v) the electron spin–electron spin interactions (exchange interactions and dipole–dipole interactions). Some of these interactions are directly observable in the CW EPR spectrum, whilst others are indirectly detected through their perturbation to the EPR signal. An understanding of these interactions and how they are manifested in the EPR spectra is very important when learning about EPR spectroscopy.

The applications of EPR spectroscopy are determined by the information that can be obtained from analysis of the EPR spectra, which is very system-dependent. EPR is undoubtedly the most powerful technique for the structural characterization of paramagnetic systems. Combined with the use of free radicals as molecular probes, it is also highly suitable for studying structural features of diamagnetic nanoscale and supramolecular structures and surfaces. Some common applications, illustrating the wealth of information that can be obtained by EPR, are listed below:

Molecular and supramolecular structure In many cases, EPR spectra make it possible to not only identify and unambiguously assign a structure to a particular radical, but also to obtain further geometrical details about the spin system. For example, detailed information about the structure of organic and inorganic radicals, including the precise location of spin-active nuclei in the vicinity of the unpaired electron (e.g. within about 0.5 nm) can be obtained.

For systems with two (or more) unpaired electrons, inter-distances up to 6–10.2 nm and angles between the two spin-bearing groups can be determined by advanced pulsed EPR techniques (see Chapter 9).

Electronic structure EPR spectroscopy provides a tool to directly probe the distribution of electron density in delocalized organic and inorganic systems. This is essential for understanding chemical and biochemical reactivity of paramagnetic compounds.

Dynamics EPR can probe diffusional or reaction dynamics on a large range of timescales, from 0.1 ns to 1 ms. This is very useful in physical chemistry, e.g. for characterization of complex systems such as colloids, gels, and liquid crystals.

Concentration EPR can be used to quantify paramagnetic species in a diverse range of samples. This has many applications, which include EPR dosimetry (Fig. 1.1b), detection and quantification of free radical intermediates in chemical reactions, measurement of Reactive Oxygen Species (ROS), which are very important in biological and medicinal studies, EPR oximetry applications (as oxygen is paramagnetic, its concentration can be measured using EPR techniques), and determination of antioxidant properties (in many biological and industrial applications, radical concentrations are controlled by the use of antioxidants, so that EPR spectroscopy is often used to assess the efficiency of the antioxidants).

1.6 Summary

- EPR is a magnetic resonance spectroscopic technique used to study systems containing unpaired electrons.
- The technique relies on the resonant absorption of the $\boldsymbol{B}_1$ field of microwave radiation by the paramagnetic sample when placed within an external magnetic field.
- The $\boldsymbol{B}_1$ field interacts with permanent magnetic dipole moments in the sample arising from unpaired electrons.
- The two spin states of the electron were originally proposed by Goudsmit and Uhlenbeck and later proven experimentally by Stern and Gerlach. The first reported observation of an EPR spectrum is credited to Zavoisky in 1945.
- The CW EPR spectrum is a derivative plot of microwave absorption versus applied magnetic field.
- Numerous magnetic interactions of the electron with its surroundings are seen directly or indirectly in the spectrum.
- The observed interactions include the electron Zeeman interactions, the zero-field splittings, the hyperfine interactions, the nuclear Zeeman and quadrupole interactions, and the electron spin-spin interactions.
- EPR provides detailed information about a paramagnetic system, including molecular and supramolecular structure, electronic structure, dynamics, and concentration.

2 Theory of continuous wave (CW) EPR spectroscopy

2.1 Introduction

In this chapter, the reader will be introduced to the theoretical background underpinning continuous wave (CW) EPR spectroscopy, focussing on the electron Zeeman, nuclear Zeeman, and hyperfine interactions. The chapter begins with a discussion of the electron spin angular momentum and considers the effect of placing the unpaired electron in an external magnetic field from a classical perspective. The influence of the hyperfine interaction and the relaxation mechanism on the EPR spectrum will also be presented.

2.2 Angular momentum and electron magnetic moment

Spin angular momentum

Electrons can be regarded as negatively charged particles possessing an intrinsic property called spin, characterized by the *spin angular momentum* $\boldsymbol{S}$. This property is a vector with a magnitude quantized in units of $\hbar$ (= $h/(2\pi)$):

$$|\boldsymbol{S}| = \sqrt{S(S+1)} \tag{2.1}$$

where S is the electron spin quantum number. In the vector representation (Fig. 2.1), the length of $|\boldsymbol{S}|$ is given by eqn 2.1. There are $2S + 1$ projections allowed onto an arbitrarily chosen axis, taken as the z axis, which coincides with the direction of the applied external magnetic field vector $\boldsymbol{B}$.

Therefore, the allowed orientations of the spin angular momentum along the z axis (also quantized in units of $\hbar$) are given by the m_S values of the spin, where m_S is the electron spin angular momentum quantum number with $2S+1$ values in integral steps between $+S$ and $-S$:

$$m_S = -S, -S+1 \ldots S-1, S \tag{2.2}$$

An electron with $S = ½$ therefore has two possible spin states, denoted as 'spin-up' (with m_S value = +½, labelled with the ↑ symbol and called the α-spin) and 'spin-down' (with m_S value = −½, labelled ↓ or β-spin). In the absence of a magnetic field, the two electron spin states have the same energy and the probability of the electron being in either spin state is equal.

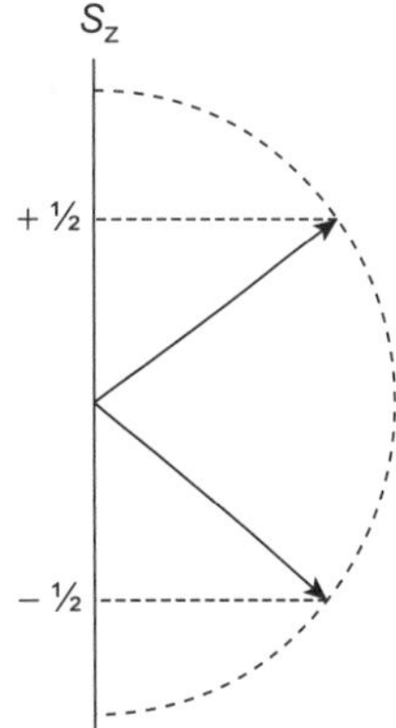

Fig. 2.1 Representation of the space quantization of the electron spin angular momentum. The magnitude of the vectors are $[S(S+1)]^{1/2}$ (eqn 2.1) in units of $\hbar$.

In quantum mechanics, the state of a system is described by a wavefunction, ψ. An operator then acts on a state to give another state. Each operator has a set of eigenvalues. For an electron spin, m_S represent the eigenvalues of the operator $\hat{S}_z$. Formally this is written as:

$$\hat{S}_z|\psi\rangle = m_S|\psi\rangle.$$

Spin magnetic moment

The *magnetogyric* or *gyromagnetic ratio* (symbol γ) of a particle is the ratio of the magnetic dipole moment to the angular momentum. $\gamma_e = -1.760859644(11) \times 10^{11}$ $s^{-1}\,T^{-1}$ and $\gamma_p = 2.675221900(18) \times 10^{8}$ $s^{-1}\,T^{-1}$.

Although spin angular momentum is quantized in units of $\hbar$ $(=h/2\pi)$, many texts omit these units in their treatment of **S**.

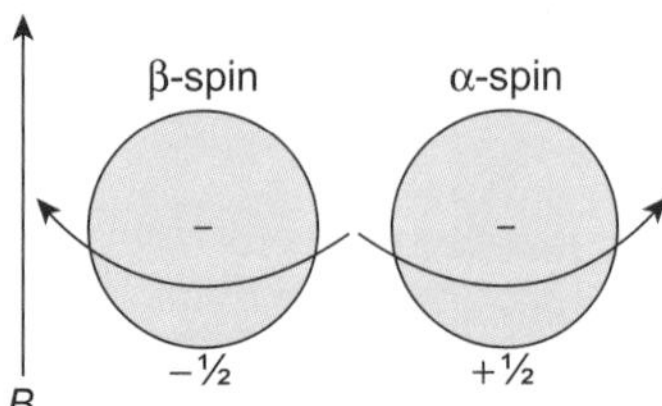

Fig. 2.2 The two orientations of the electron magnetic dipole moments, $\boldsymbol{\mu}_S$, arising from the β- and α-spin states, aligned parallel (clockwise) or anti-parallel (anticlockwise) to the applied field (*z*-direction).

In classical terms, it is convenient to consider the electron as a particle of mass m_e and charge *e* rotating about an axis with spin angular momentum **S** which produces a current. This circulating current generates a magnetic dipole moment (μ_S) related to the gyromagnetic ratio (γ) by:

$$\mu_S = -\gamma \boldsymbol{S}\hbar = -g_e \frac{e\hbar}{2m_e}\boldsymbol{S} \tag{2.3}$$

In this equation, g_e is the free-electron *g*-factor (with an approximate value of 2.0023), a dimensionless factor that corrects the magnetic moment of the quantum electron from the classical result. This eqn 2.3 is more commonly expressed in terms of the Bohr magneton ($\mu_B = e\hbar/(2m_e) = 9.274 \times 10^{-24}\,J\,T^{-1}$):

$$\mu_S = -g_e\mu_B\boldsymbol{S} \tag{2.4}$$

The *z*-component of the magnetic moment (μ_z) along the applied magnetic field **B** direction can be related to the m_S electron spin states, resulting in two electron magnetic dipole moments (Fig. 2.2) given by:

$$\mu_z = -g_e\mu_B m_S \tag{2.5}$$

The negative sign in eqns 2.3–2.5 arises due to the negative charge on the electron (and the assignment of g_e and μ_B as positive values). This indicates that the electron magnetic moment is collinear and anti-parallel to the direction of the spin angular momentum.

Influence of magnetic fields and microwave fields

B is defined as the magnetic field induction *vector*, while *B* is the magnetic field induction *magnitude*. Both have units of Tesla (T) or Gauss (G); 1 G = 0.1 mT.

In older texts, the magnetic field is defined by ***H***, with different dimensions and units to ***B***. The two are related by $\boldsymbol{H} = \boldsymbol{B}/(\kappa_m\mu_0)$ (κ_m = dimensionless parameter = 1 for a vacuum, μ_0 = permeability of a vacuum).

In the absence of a magnetic field the two electron spin states are degenerate. This degeneracy is removed in the presence of an external magnetic field. In classical terms, the interaction energy (*E*) of the electron magnetic moment with **B** is given by:

$$E = -\mu_S \cdot \boldsymbol{B} \tag{2.6}$$

From eqn 2.6 it can be understood that *E* is directly proportional to $|\boldsymbol{B}|$. Taking the space quantization axis to be aligned along the external magnetic field, the scalar value μ_z (i.e. the projection of μ_S onto **B**) in a magnetic field of strength *B* (i.e. the magnitude of the vector **B**) results in:

$$E = -\mu_z B \tag{2.7}$$

Substituting the expression for the magnetic dipole moment given in eqn 2.5 results in the following energy term:

$$E = g_e\mu_B m_S \times B \tag{2.8}$$

The *Zeeman effect* describes the splitting of the electron spin energy levels in a magnetic field.

The *Zeeman interaction* describes the electron magnetic moment interacting with the applied field.

The two components of the magnetic dipole moment along the *z*-direction (i.e. the solutions to eqn 2.5) give two states of different energy, referred to as the *electron Zeeman levels* (Fig. 2.3). The lowest energy β-spin state, labelled E_1', corresponds to the electron magnetic moment aligned parallel to the applied field and the high energy state arises due to the anti-parallel alignment (the α-spin state, labelled E_2').

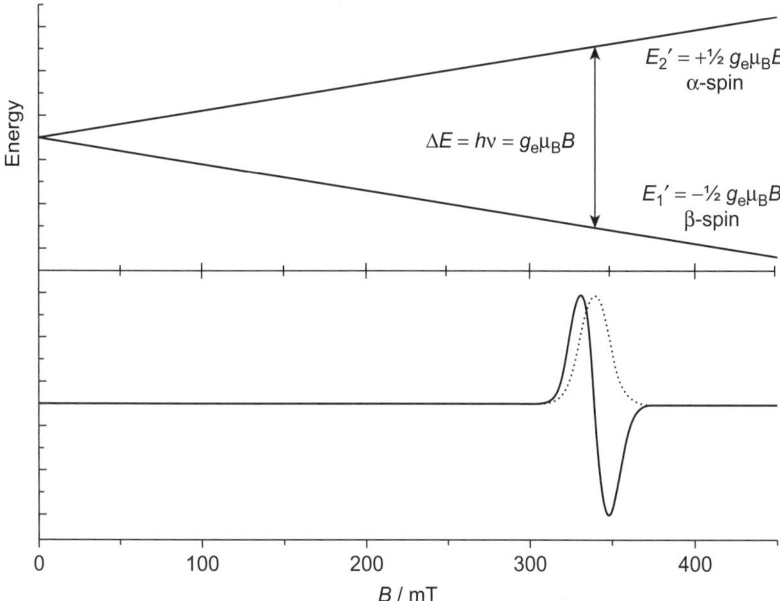

Fig. 2.3 The electron Zeeman levels for an unpaired electron in an external magnetic field of increasing magnitude B (in units of mT). Resonant energy absorption (eqn 2.9) leads to an electron spin 'flip' or transition resulting in an EPR signal. This signal can be presented in absorption (dotted) or first derivative (solid) mode.

For $g \approx 2$, the application of magnetic fields with intensities between 0.1 and 1.5 T results in resonance frequencies in the microwave regime, between 2.8 and 42 GHz (see Chapter 3).

In EPR spectroscopy, a sample containing unpaired electrons is therefore irradiated with electromagnetic radiation ($h\nu$) in the presence of an external field. When the energy difference (ΔE) created by the applied field, between the two levels E_1' and E_2', matches the available energy quantum ($h\nu$) resonance, absorption occurs leading to an EPR resonance, as indicated in Fig. 2.3. As discussed in more detail in Chapter 3, continuous wave (CW) mode EPR spectra are recorded in first derivative mode.

The selection rule in EPR spectroscopy is $\Delta m_S = \pm 1$, to conserve angular momentum, so that the resulting resonance condition may be written as:

$$\Delta E = h\nu = E_2' - E_1' = g_e\mu_B B \tag{2.9}$$

where ν is the frequency of the applied microwave radiation. Typically, in a CW EPR experiment, ν is held constant while B is incremented.

The magnetic field experienced by the electron in a molecule will differ from $\boldsymbol{B}$, such that the exact resonance frequency of the EPR signal is characteristic of the chemical environment. The presence of local fields at the electron add to the externally applied field $\boldsymbol{B}$ to give an effective field experienced by the electron:

$$\boldsymbol{B}_{\text{eff}} = \boldsymbol{B} + \boldsymbol{B}_{\text{local}} \tag{2.10}$$

$\boldsymbol{B}_{\text{eff}}$ should replace $\boldsymbol{B}$ when considering the interaction energy of an electron in an applied field, although in practice it is more convenient to retain $\boldsymbol{B}$ (which can easily be measured) and replace g_e with an *effective* g factor:

$$\boldsymbol{B}_{\text{eff}} = (g/g_e)\boldsymbol{B} \tag{2.11}$$

It should be noted that the local field present at the electron may also arise from the orbital angular momentum (***L***), in addition to the spin angular momentum which was only considered in eqn 2.8. In general $L = 0$ for electronic ground states of molecules that are non-degenerate, as the local molecular environment can lift the degeneracy. However, mixing of ground and excited states can cause the electron spin and orbital angular momentum to couple, and this may cause the g values to deviate considerably from g_e. The effects of *spin-orbit coupling* on g are considered in Chapter 5. Furthermore, when these $\boldsymbol{B}_{local}$ fields are not collinear with ***B***, the g value in eqn 2.11 becomes anisotropic and must be replaced by a 3×3 **g** matrix (see Chapter 5). For the moment it is simply important to appreciate that the effective g value in eqn 2.11 reflects any induced local magnetic fields that occur in real paramagnetic systems.

Table 2.1 Standard reference markers used in EPR.

Standard	g-value
DPPH*	2.0036
Weak Pitch	2.0028
Cr^{3+} (MgO matrix)	1.97989

*DPPH = *2,2-diphenyl-1-picrylhydrazyl* (structure shown in Chapter 4).

2.3 Measurement of the g value

From the previous discussion, one can see that the key parameter of interest is the g value. This determines the resonant field position and, as suggested in the previous paragraph, is influenced by the chemical environment of the electron. For systems containing atoms with small atomic masses (e.g. organic radicals, see Chapter 4), the extent of the spin-orbit coupling is small and the resulting g values are usually close to free spin ($g_e \approx 2.0023$). However, deviations from g_e are observed for samples with large orbital angular momenta contributions, such as main group radicals and transition metal ions (see Chapter 6).

Direct measurement: $g = h\nu/(\mu_B B)$.

Using a reference: $g = g_{ref}B_{ref}/B$. This equation follows from eqn 2.9 if the frequency is kept constant, which is the case in CW EPR measurements.

The accurate measurement of the g value is extremely important in EPR spectroscopy, and can be used to identify the radical. Two methods are available to extract the g values experimentally, using either direct measurement of the resonance field and applied microwave frequency or by referencing an unknown sample (see Table 2.1) to a standard with a known g value. The reference (marker) sample may be placed in the resonator with the sample of interest so that they both experience the same external magnetic field. Alternatively, double (coupled) cavities are available that ensure both samples are measured under identical conditions.

The reference standard should, if possible, be chosen such that the signal from the standard does not overlap with any resonances from the unknown sample. For example, in Fig. 2.4, the EPR signal of the DPPH reference standard is observed at $B_{ref} = 338.75$ mT, and does not overlap with the resonance from the unknown sample.

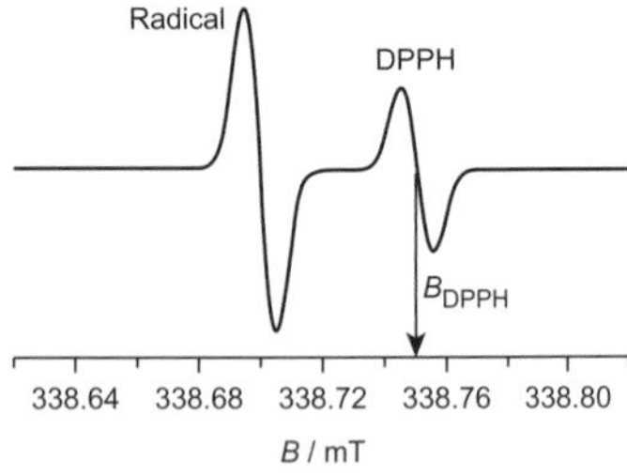

Fig. 2.4 The g value can be determined through direct measurement or with a reference standard.

2.4 The nuclear spin and hyperfine interaction

If the interaction of the unpaired electron with an applied magnetic field was the only effect detectable by EPR spectroscopy, then the spectra would be of limited interest. However, the presence of nuclei with a magnetic spin (symbol I) adds further magnetic interactions to the electron spin system, which results in multi-line EPR spectra that contain a wealth of information. The interaction of the nuclear spin with ***B*** results in a nuclear Zeeman splitting, and the resulting interaction between the electron and nuclear magnetic moments is described by the hyperfine interaction. These interactions give rise to additional terms in eqn 2.8, creating small but significant perturbations to the electron spin energies (E), which are summarized as:

In eqn 2.12, only the isotropic g and a values are considered; anisotropy (see Chapter 5) is neglected.

$$E = g\mu_B B m_S - g_N \mu_N B m_I + a m_S m_I \tag{2.12}$$

These two additional terms in eqn 2.12 are referred to as the nuclear Zeeman splitting and the hyperfine interaction, respectively.

Equation 2.12 describes the energy of the spin system in a magnetic field. In quantum mechanics, the energy is defined using the Hamiltonian ($\hat{H}$) and is expressed as:

$$\hat{H} = g\mu_B B\hat{S}_z - g_N\mu_N B\hat{I}_z + a\hat{S}_z\hat{I}_z$$

where $\hat{S}_z$ and $\hat{I}_z$ are the z-components of the electron and nuclear spin angular momentum operators. This is treated in more detail in Chapter 5.

The nuclear Zeeman splitting

In section 2.2, the energy of the two electron spin states was shown to be related to the orientation of the magnetic dipole moment in the external magnetic field $\boldsymbol{B}$, leading to two electron Zeeman levels. In analogy with $\boldsymbol{\mu}_S$ (eqn 2.4), nuclei that possess a non-zero nuclear spin quantum number, $\boldsymbol{I}$, will also have an associated magnetic moment $\boldsymbol{\mu}_I$, written as:

$$\mu_I = g_N\mu_N \boldsymbol{I} \qquad (2.13)$$

where g_N is the effective nuclear g-factor (which can have positive or negative values), μ_N is the nuclear magneton ($= e\hbar/2m_p = 5.0508 \times 10^{-27}$ J T^{-1}), and m_p is the mass of a proton. Similar to the spin angular momentum (eqn 2.1), the magnitude of the nuclear spin moments are also quantized in units of $\hbar$ and given as:

$$|\boldsymbol{I}| = \sqrt{I(I+1)} \qquad (2.14)$$

As a result, the vector $\boldsymbol{I}$ can assume $2I + 1$ discrete orientations which are given by the magnetic quantum number (m_I):

$$m_I = -I, -I+1 \ldots I-1, I \qquad (2.15)$$

By analogy to the electron spin (eqn 2.5), the z-component of the nuclear magnetic dipole moment is related to the m_I nuclear spin states by:

$$\mu_z = g_N\mu_N m_I \qquad (2.16)$$

Protons can have an orbital and spin angular momentum, similar to electrons. Neutrons have a spin only contribution. As a result, the contribution to g_N from the spin part of the proton and neutron have opposite sign. This is the origin of the differing signs for g_N with different numbers of nucleons.

The nuclear spin angular momentum is formed by coupling the angular momenta of the nucleons (i.e. protons and neutrons). Isotopes of the same element may therefore have different nuclear spin quantum numbers (for example, see Table 2.2). In the absence of a magnetic field, the permitted z-components of nuclei with $I \geq ½$ (using the same space quantization as illustrated in Fig. 2.1) have the same energy. In analogy with the electron magnetic moment, the degeneracy of the $2I + 1$ nuclear energy levels is removed by the influence of a magnetic field. The energy of these nuclear Zeeman levels is given by:

Table 2.2 Nuclear spin quantum numbers (I) of some commonly occurring nuclides.

I	Nuclide
0	^{12}C, ^{16}O
½	^{1}H, ^{13}C, ^{15}N, ^{19}F, ^{29}Si, ^{31}P
1	^{2}H, ^{14}N
3/2	^{11}B, ^{35,37}Cl, ^{63,65}Cu
5/2	^{17}O, ^{55}Mn
7/2	^{51}V

$$E = g_N\mu_N B m_I \qquad (2.17)$$

It is important to note that g_N can be positive (e.g. for ^{14}N) or negative (e.g. for ^{15}N). As a result, the magnetic dipole moments of the nuclei will align with or against the applied field. By comparison to the electron Zeeman levels, the $2I + 1$ nuclear Zeeman energy levels are equally spaced with an energy gap given by γB, which due to the much smaller μ_N value is considerably smaller than the electron Zeeman separation.

The additional energy levels arising from the nuclear Zeeman interaction are illustrated in Fig. 2.5, for a two-spin system ($S = I = ½$). Each of the non-degenerate electron Zeeman levels is split into $2I + 1$ nuclear Zeeman levels. For an $I = ½$ system, this results in four discrete energy levels (labelled E_1 to E_4) due to the combined interaction of the electron and nuclear magnetic dipole moments with the external magnetic field (two nuclear Zeeman levels in each electron spin manifold).

Let us assume positive signs for electron and nuclear g factors (g_e and g_N). As the electron Zeeman term in eqn 2.12 is positive, but the nuclear Zeeman term is negative, an electron in the spin-down state ($m_S = -½$) is lower in energy than in the spin-up state ($m_S = +½$), but a nucleus in the spin-down state ($m_I = -½$) is higher in energy than in the spin-up state ($m_I = +½$).

The separation between the two resonance lines in Fig. 2.5 (labelled EPR I and EPR II) gives a in magnetic field units (mT or G). However, the hyperfine is an energy and therefore a (in field units) should be converted to frequency or wavenumber units (see Appendix A).

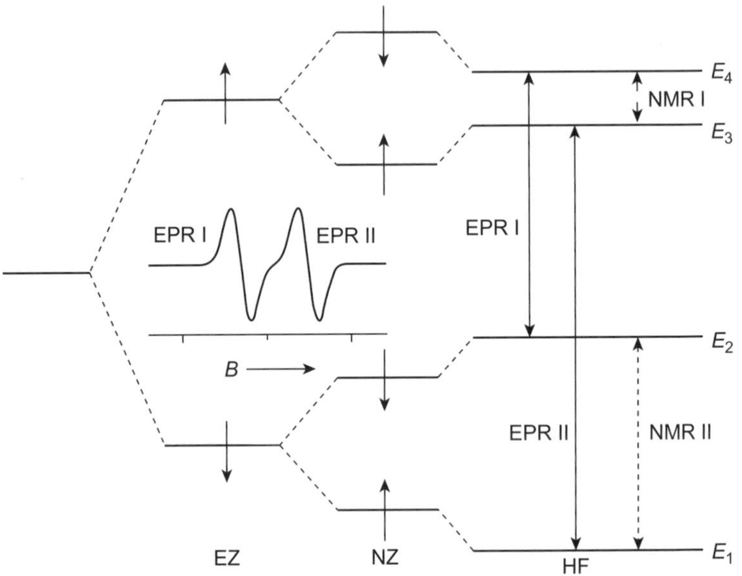

Fig. 2.5 Energy level diagram in a fixed magnetic field for an $S = I = ½$, spin system in the high-field approximation, showing the electron Zeeman (EZ) and nuclear Zeeman (NZ) levels, and the perturbation arising from the hyperfine interaction (HF). In this case, g_N and a are positive, and $a < g_N\mu_N B$. The two allowed EPR transitions (solid arrows) result in the experimentally observed resonances labelled EPR I and EPR II (shown in the inset).

The hyperfine interaction

The magnitude of the magnetic interaction between the electron and nuclear spin can be described via the *hyperfine splitting constant* or the *hyperfine coupling constant*. This is expanded in more detail in Chapters 4 and 5.

The final term in eqn 2.12, am_Sm_I, describes the hyperfine interaction that arises from the interaction between the electron and nuclear magnetic dipole moments with each other. The effect of the hyperfine interaction is to create a *perturbation* of the nuclear Zeeman energy levels towards higher or lower energy. The extent of this perturbation is reflected in the magnitude of the interaction as described by the hyperfine splitting a.

The solutions to eqn 2.12 for the corresponding energies of the two-spin system ($m_S = \pm½$ and $m_I = \pm½$) are given by:

In the high-field approximation:

$$|a| << g\mu_B B$$

Let us assume a positive sign for the hyperfine constant a. As the hyperfine interaction term in eqn 2.12 includes a product of m_S and m_I values, this interaction reduces the energy of the state if m_S and m_I have opposite signs (e.g. $m_S = -½$ and $m_I = +½$ for the level E_1), but increases the energy of the state if m_S and m_I have the same sign (e.g. $m_S = +½$ and $m_I = +½$ for level E_3, Fig. 2.5).

$$E_1 = -\tfrac{1}{2}g_e\mu_B \cdot B - \tfrac{1}{2}g_N\mu_N \cdot B - \tfrac{1}{4}a \tag{2.18a}$$

$$E_2 = -\tfrac{1}{2}g_e\mu_B \cdot B + \tfrac{1}{2}g_N\mu_N \cdot B + \tfrac{1}{4}a \tag{2.18b}$$

$$E_3 = +\tfrac{1}{2}g_e\mu_B \cdot B - \tfrac{1}{2}g_N\mu_N \cdot B + \tfrac{1}{4}a \tag{2.18c}$$

$$E_4 = +\tfrac{1}{2}g_e\mu_B \cdot B + \tfrac{1}{2}g_N\mu_N \cdot B - \tfrac{1}{4}a \tag{2.18d}$$

For the case of positive g_N and a, this results in the energy level diagram shown in Fig. 2.5. The order of the m_I levels within each nuclear Zeeman manifold is determined by the sign of g_N (e.g. for $g_N > 0$, the $m_I = +½$ state lies lowest in energy within the $m_S = -½$ state).

In CW EPR spectroscopy, the transitions from one state to another which occur with the highest probabilities are those corresponding to $\Delta m_S = \pm1$ and $\Delta m_I = 0$. That is, the electron spin state must change following the transition but the nuclear spin state must remain the same. Therefore, for the two-spin system shown in Fig. 2.5, two allowed EPR transitions occur labelled EPR I (i.e. $E_2 \rightarrow E_4$) and EPR II ($E_1 \rightarrow E_3$), indicated with solid arrows. The different arrow lengths for EPR I and EPR II represent different energy quanta that are required to induce the transitions, resulting in two resonance lines in the EPR spectrum (inset Fig. 2.5). The magnitude of the separation between these two resonance lines is given by the hyperfine splitting constant:

$$|E_1 \rightarrow E_3| - |E_2 \rightarrow E_4| = a \tag{2.19}$$

The resonance frequencies of the two transitions shown in Fig. 2.5 are given by: $h\nu = g\mu_B B_{res} \pm \frac{1}{2}a$. In general the resonance frequency can be written $h\nu = g\mu_B B_{res} + m_I a$ while the resonant fields are given as: $B_{res} = \frac{h\nu}{g\mu_B} - m_I\left(\frac{a}{g\mu_B}\right)$. The two lines in the EPR spectrum (Fig. 2.5) are thus separated by $a/(g\mu_B)$ if a is expressed in energy units and centred on $B = h\nu/g\mu_B$.

There are two contributions to the electron spin–nuclear spin (i.e. hyperfine) interaction that arise from the regions of space inside and outside the nuclear volume. These are referred to as the isotropic (a) and anisotropic or dipolar interactions (T), which are presented in the following narrative.

μ_0 is the vacuum permeability with a value of $4\pi \times 10^{-7} = 12.566370614 \times 10^{-7}$ $T^2 J^{-1} m^3$

The isotropic (Fermi contact) hyperfine interaction

From a quantum mechanics consideration, there is a small probability that the electron may enter the nuclear volume. Inside the nucleus, the hyperfine field originating from the nuclear magnetic dipole is constant in all directions. This hyperfine interaction is called the *isotropic interaction*, or Fermi contact interaction, and is a direct measure of the interaction between the electron and nuclear magnetic dipole moments as a result of the finite probability that the unpaired electron will be located at the nucleus. It arises exclusively for s-orbitals (or orbitals with partial s-character) (Fig. 2.6). The isotropic coupling constant, a_0, in energy units, is given by:

$$a_0 = (2\mu_0/3) g\mu_B g_N \mu_N |\psi(0)|^2 \tag{2.20}$$

where $|\psi(0)|^2$ is the square of the absolute value of the unpaired electron wavefunction evaluated at the nucleus. The magnitude of the experimentally observed isotropic hyperfine coupling can thus be related to the electron spin density located at the nucleus, through the theoretically determined isotropic hyperfine coupling values. Isotropic couplings are observed in fluid solution EPR spectra, which are discussed in detail in Chapter 4.

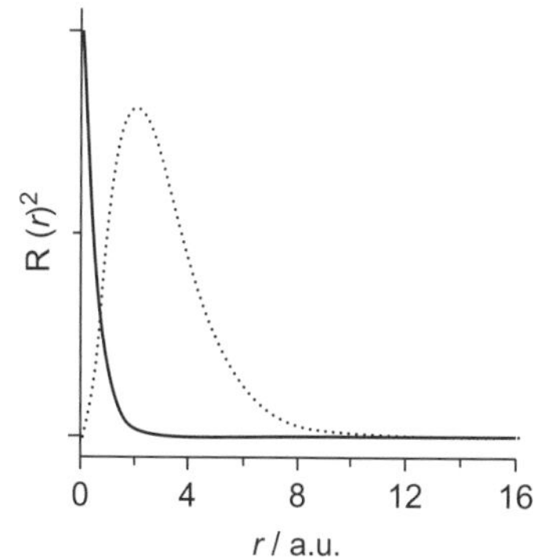

Fig. 2.6 Radial wavefunctions of 1s (solid) and 2p (dotted) orbitals, indicating the probability density of the electron.

In EPR spectroscopy, the sign of the isotropic hyperfine coupling constant is very important even though the absolute sign ($\pm a$) cannot be determined directly from the spectrum. The sign will affect the relative ordering of the energy levels as presented in Fig. 2.5. According to eqn 2.20, the sign of g_N directly determines the sign of a. In other words, the sign of a will dictate the parallel or antiparallel orientation of the magnetic moments of the electron and nucleus. Furthermore, in multi-electron systems, an outer unpaired electron spin may polarize inner electron pairs to adopt a parallel or antiparallel alignment. This spin polarization effect (expanded further in Chapter 4) can also indirectly influence the sign of a.

The anisotropic (dipolar) hyperfine interaction

The second contribution to the hyperfine interaction arises from a classical dipole-dipole interaction, due to the hyperfine field in the region external to the nucleus. The energy of this anisotropic, or dipolar, hyperfine interaction (E_{dip}) is dependent on the relative orientation and distance between the electron and nuclear magnetic dipole moments, as described by the classical energy term:

$$E_{dip} = (\mu_0 / 4\pi)[\mu_S^T.\mu_I / r^3 - 3(\mu_S^T.\boldsymbol{r})(\mu_I^T.\boldsymbol{r}) / r^5] \tag{2.21}$$

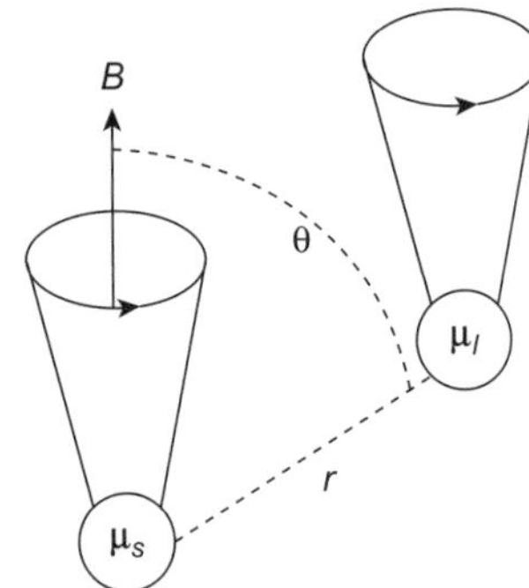

Fig. 2.7 The classical dipolar interaction between $\boldsymbol{\mu}_S$ and $\boldsymbol{\mu}_I$ is a function of the distance r and their relative orientation θ (eqn 2.21).

where $\boldsymbol{r}$ is the vector relating the electron position in a coordinate frame centred on the nucleus, and r is the electron–nucleus distance (Fig. 2.7).

Equation 2.21 must be averaged over the entire probability of the spin distribution, therefore E_{dip} is averaged to zero in a spherical electron cloud (e.g. in an s-orbital), or for rapidly tumbling molecules where the anisotropy with respect to the external field is motionally averaged, and comes to a finite value in the case of axially symmetric orbitals (e.g. p-orbitals). Spectra arising from anisotropic interactions (as observed in the solid state) are discussed in Chapter 5.

The nuclear quadrupole interaction

Many nuclei have $I > ½$ (see Table 2.2). In such cases, a non-spherical distribution of charge is present at the nucleus and this generates a quadrupole moment (P), which interacts with electric field gradients at the nucleus. This interaction is rarely observed directly in EPR spectra, creating only small second-order perturbations to the electron spin energy levels. However, in some advanced EPR techniques (see Chapter 9), these quadrupole interactions can be dominant and appear as first-order perturbations. This interaction is treated in greater detail in Chapter 5.

2.5 Spin relaxation mechanisms

Bloch model

In a real macroscopic system containing many electron spin centres, the electron spins will interact with each other and the surrounding lattice. These interactions result in relaxation mechanisms that act to restore the system to equilibrium (e.g. after spin flipping brought about by the absorption of microwave energy during the recording of the EPR spectrum). A valuable tool for discussion of relaxation mechanisms is provided by the *Bloch model*. The actual quantity detected in an EPR experiment is the macroscopic magnetization vector, $\boldsymbol{M}$ (with components of M_x, M_y, and M_z), defined as the net magnetic moment per unit volume (V) due to the macroscopic collection of magnetic dipole moments in the sample:

$$\boldsymbol{M} = \frac{1}{V}\sum \mu_S \tag{2.22}$$

Using the convention of aligning $\boldsymbol{B}$ along the z-direction, if the system remains in thermal equilibrium with the lattice, the resulting magnetization will be directed along the z axis due to the excess of spins in the lower energy β-spin state. If the relative populations of the α- and β-spin states are changed by the absorption of microwave energy, the bulk magnetization is forced out of equilibrium. The time evolution of the total spin magnetization vector is described in the presence of static (and oscillatory) magnetic fields as a function of the spin–lattice and spin–spin relaxation mechanisms. The Bloch equations for the components of the bulk magnetization are:

$$\frac{dM_x}{dt} = \gamma_e B M_y - \frac{M_x}{T_2} \tag{2.23a}$$

$$\frac{dM_y}{dt} = -\gamma_e B M_x - \frac{M_y}{T_2} \tag{2.23b}$$

$$\frac{dM_z}{dt} = \frac{M_z^0 - M_z}{T_1} \tag{2.23c}$$

From eqns 2.23a and b it is seen that the time evolution of the M_x and M_y components is characterized by the spin-spin (or transverse) relaxation time T_2, which relates energy exchange processes *within* the spin system (and does not involve the lattice). Deviation of the *z* component of the magnetization M_z from the equilibrium value M_z^0, is restored through spin-lattice interactions. The M_z component relaxes with the spin-lattice (or longitudinal) relaxation time T_1, which characterizes energy transfer *to or from* the lattice. Generally, T_1 and T_2 are of different magnitude and can be determined directly using advanced Fourier transform methods (see Chapter 9). At equilibrium, the spin temperature T_S of the sample is given by:

$$\frac{N_\alpha}{N_\beta} = \exp\left(-\frac{\Delta E}{kT_S}\right) \tag{2.24}$$

where N_α and N_β are the occupancies of the α- and β-electron spin manifolds, ΔE is the energy separation between them, and *k* is the Boltzmann constant. Owing to the small energy gap, ΔE, a slight excess of spins exists in the lowest energy β state (Fig. 2.8a). Absorption of a suitable quantum of microwave energy induces a transition of some electron spins between the two Zeeman levels (Fig. 2.8b), such that the ratio N_α/N_β is altered. When the population of the two spin states is equal (Fig. 2.8c), a condition referred to as saturation is reached. Under these conditions, no transitions between states can be induced by further absorption of microwave energy, which leads to a decrease in the amplitude (and ultimately disappearance) of the associated EPR signal. A useful parameter used in EPR is the polarization, *P*, which measures the excess of spins in the β state and is given by:

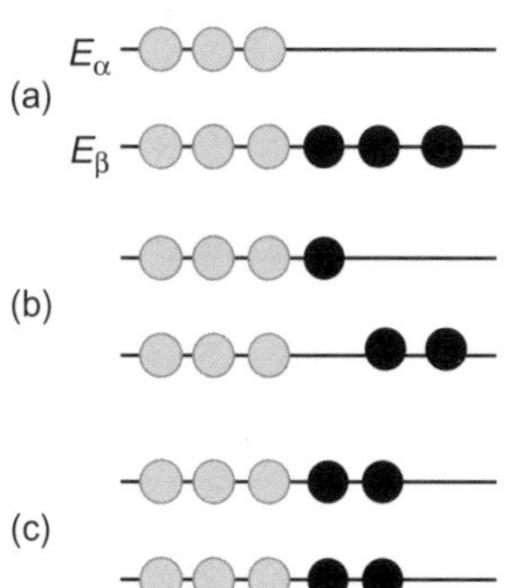

Fig. 2.8 Two-level spin system at temperature *T*: (a) at thermal equilibrium, $T_S = T$; (b) some electrons excited from the ground state, $T_S > T$; (c) equal population of the two energy levels, $N_\alpha = N_\beta$.

$$P = \frac{N_\alpha - N_\beta}{N_\alpha + N_\beta} \approx \frac{h\gamma B}{4\pi kT} \tag{2.25}$$

From eqn 2.25 it can be understood that decreasing the sample temperature results in increased polarization, which can improve the signal quality in the resulting EPR spectrum.

2.6 Summary

- An electron has a characteristic spin quantum number, *S*, with a characteristic magnetic moment, μ_S.
- There are $2S + 1$ quantized energy levels for the electron.
- The degeneracy of the two electron spin states is lifted in the presence of a magnetic field, by an amount proportional to μ_S and to the strength of the applied field ***B***.
- Conventional EPR spectroscopy uses microwave radiation to induce transitions between energy levels.

- The exact resonance frequency of the EPR signal is characteristic of the chemical environment.
- Coupling of the electron nuclear spins results in hyperfine interactions.

2.7 Exercises

2.1) Calculate the g value of the resonance due to the unknown radical in Fig. 2.4. Assume a microwave frequency of 9.5 GHz and $g_{DPPH} = 2.0036$.

2.2) Draw the energy level diagram for an unpaired electron interacting with a single nucleus of i) ^{14}N, ii) ^{11}B, and iii) ^{15}N. Assume the magnitude of the hyperfine interaction is much larger than the energy associated with the nuclear Zeeman interaction (i.e. $|a| >> g_N \mu_N B \cdot m_I$).

2.3) In a magnetic field of 335 mT, calculate the polarization parameters for the operating temperatures of 4, 50, and 298 K.

2.4) The isotropic hyperfine coupling of a ^{1}H atom is 50.69 mT. For an observed hyperfine coupling of 7.1 mT, calculate the electron spin density on a ^{1}H nucleus.

2.5) Using the classical expression for a magnetic moment ($\mu = IA$, where I is the effective current and A is the area of a circular orbit), show that the orbital magnetic moment of an electron can be described by:

$$\mu_L = \frac{e\hbar}{2m_e}\sqrt{l(l+1)}$$

3 Experimental methods in CW EPR

3.1 Introduction

As described in Chapter 2, the fundamental basis of EPR spectroscopy involves the observation of electron spin transitions in the presence of a magnetic field. In the continuous wave (abbreviated CW) technique, a continuous source of microwave (MW) radiation of fixed frequency is applied to the sample which is mounted in a cavity or resonator to induce the spin transitions. Detection of the absorbed MW energy, as a function of the magnetic field, results in the observed EPR signal. This chapter discusses the essential hardware components required to perform a CW EPR experiment, including the microwave bridge, the resonant cavity, the external magnet, and the console (Fig. 3.1). The different functions of these components will be described and advice on how to optimize the EPR signal, through judicious selection of the instrumental parameters, will be given.

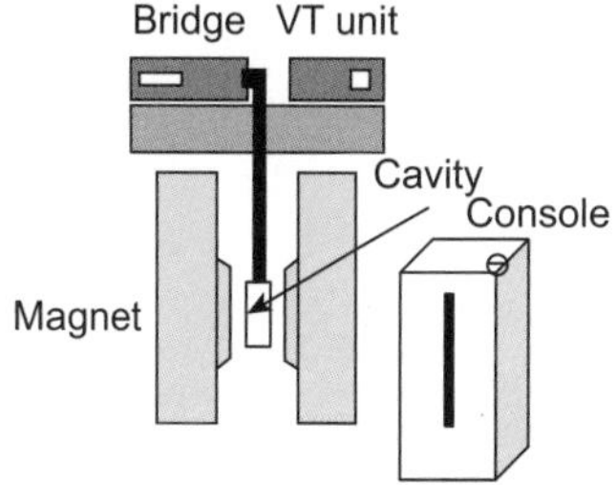

Fig. 3.1 The essential hardware components of a CW EPR spectrometer. VT unit = variable temperature unit for recording EPR spectra at the desired temperature.

3.2 Hardware components

The microwave bridge

In EPR, it is common practice to classify the microwave frequencies used for measurement into bands (Table 3.1). By far the most commonly used frequency band in EPR spectroscopy is X-band. The microwave radiation can be produced by either a klystron or a Gunn diode, before it is channelled into the microwave resonator along a waveguide or coaxial cable. The main components of the microwave bridge are highlighted in Fig. 3.2.

The isolator is located immediately after the microwave source and strongly attenuates any reflections from the system back towards the source, which would result in large fluctuations in the microwave frequency. The microwave power from the source is then passed through a directional coupler which splits the microwave power into the two paths, directed towards the resonator and a reference arm. The reference arm consists of a variable attenuator that brings the microwave detector into its linear response regime, which ensures that both large and small amplitude signals are detected with the correct amplitude ratios. The phase shifter of the reference arm sets a defined phase relationship between the reference and the reflected signal, facilitating phase-sensitive detection. The circulator then directs the incident microwave power to the resonator and the reflected power (after absorption by the sample) towards the detector.

Table 3.1 Common microwave frequency bands used in EPR spectroscopy.

Band	[a]Typical frequency/GHz	[b]Field/T	[c]Waveguide dimension/mm
L	1	0.036	196 × 98
S	3.5	0.13	72 × 34
X	9.5	0.34	23 × 10
Q	34	1.21	5.7 × 2.8
W	95	3.4	2.5 × 1.3

[a]Within each band, a range of frequencies is possible. The typical frequency is given here. [b]Field quoted in units of Tesla for $g = 2.0023$. [c]Approximate dimensions.

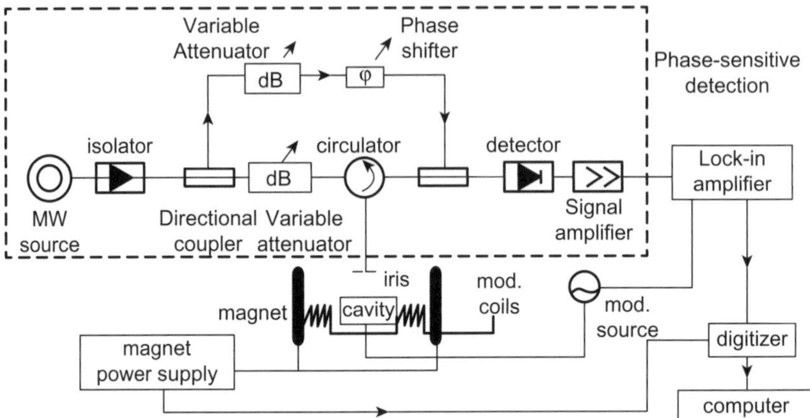

Fig. 3.2 Block diagram representing a typical CW EPR spectrometer, employing 100 kHz field modulation and phase-sensitive detection. The MW bridge components are indicated within the area of the dashed box.

The magnet

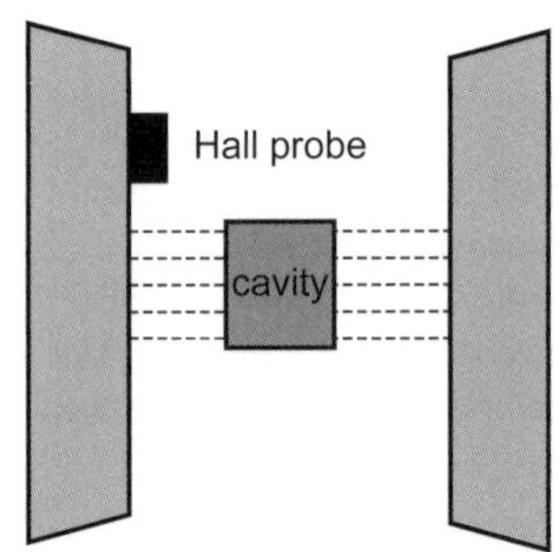

Fig. 3.3 The magnetic field is measured and regulated by a *Hall probe*, positioned slightly offset from the sample. The field is homogeneous across the sample volume.

For electromagnets (and some cryomagnets), the magnetic field is perpendicular to the sample tube axis.

The magnet assembly includes the magnet, a dedicated power supply, and a field sensor or regulator, such as a *Hall probe*. Two types of magnet systems are available and used in EPR spectroscopy, including i) electromagnets, which are capable of generating fields up to 1.5 T (i.e. suitable for measurements up to Q-band frequency), and ii) superconducting magnets, which are used for higher frequency operation (W-band and higher field measurements, see Table 3.1).

The required magnetic field range is determined by the operating microwave frequency, which is selected based on the nature of the sample under investigation. The key features of the magnet system include a homogeneous field across the entire sample volume (in which the longest axis of the homogeneous magnetic field volume points along the sample tube axis), a linear sweep across the field range of the measurement (which can span >1000 mT at high frequencies), and high stability at static field. These properties of the magnet ensure that precise g values can be determined and allow for accurate analysis of the lineshapes, linewidths, and hyperfine couplings. The magnetic field values are determined by the *Hall probe* which is positioned slightly offset from the sample within the magnet poles (Fig. 3.3). This positional displacement results in a small field offset (range 0.01–0.5 mT) and is often compensated by the instrument software. If the extent of the offset is unknown, it may be more appropriate to compare the sample to a reference standard when determining the precise g values (see section 2.3).

The microwave resonator

The microwave resonator, or cavity, is designed to enhance the microwave magnetic field ($\boldsymbol{B}_1$) at the sample (relative to the free-space value) in order to induce the EPR transitions (see Chapter 1). Following resonance absorption of microwave energy by the sample, the resonator should reflect microwave power back to the detector to enable measurement of the EPR signal. At the resonance frequency of the cavity (ν), a standing microwave is created such that in the absence of the sample all of the microwaves are contained inside the cavity, with none reflected back to the detector.

The efficiency of a resonator or cavity in storing this microwave energy is given by the quality, or Q-factor, which can be expressed as:

$$Q = \frac{\nu}{\Delta\nu} = \frac{2\pi(\text{energy stored})}{\text{energy dissipated}} \tag{3.1}$$

where $\Delta\nu$ is the resonator bandwidth, defined as the width at half height of the reflection resonance (Fig. 3.4). Energy dissipated to the walls of the cavity, through the generation of electrical currents (and subsequently heat), decreases the sensitivity that can be achieved in the measurement.

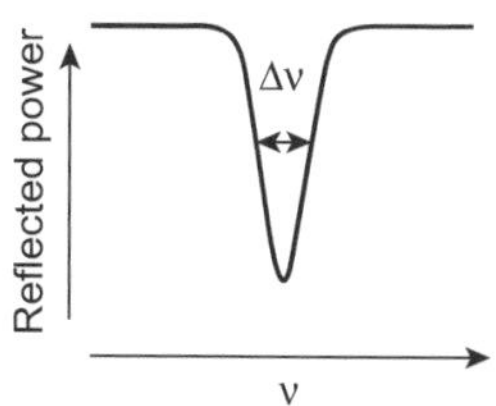

Fig. 3.4 The Q-factor is defined as the ratio between the resonator frequency, ν, and the bandwidth, $\Delta\nu$.

The process of tuning the cavity involves maximizing the Q-factor through matching the microwave resonance profile to the sample. The amount of microwaves entering and leaving the cavity is controlled by the *iris screw* (Fig. 3.2), which covers a small hole between the end of the waveguide and the cavity. At a unique position of the iris screw (which differs for each sample and operating temperature), the cavity is critically coupled such that full microwave power enters the cavity and no radiation is reflected out. During an EPR measurement, the sample absorbs some of the incoming microwave energy that modifies the coupling profile within the cavity. The effect is such that a small amount of microwave power is reflected back to the detector and this is detected as an EPR signal. The strength of the microwave field $\boldsymbol{B}_1$ generated in the resonator by the applied microwave power (P) is:

The complete absence of reflected radiation during critical coupling is often not achieved. In practice, the point of minimal reflection is observed.

$$B_1 = c \times (Q \times P)^{1/2} \tag{3.2}$$

where c is a conversion factor specific to the choice of resonator. Smaller resonators have larger conversion factors, but can only accommodate smaller sample volumes.

The interaction of the sample with the electric field component ($\boldsymbol{E}_1$) of the electromagnetic microwave radiation can lead to unwanted non-resonant absorption with electric dipole moments in the sample (particularly for highly polar solvents or materials). This may cause difficulties when tuning the resonator. Maximum sensitivity is achieved by positioning the sample in the location of maximum $\boldsymbol{B}_1$ field (corresponding to a minimum of the $\boldsymbol{E}_1$ field).

The physical dimensions of a waveguide and cavity match the microwave wavelengths (Table 3.1); higher operating frequencies require resonators with smaller dimensions. For example, the standard and commonly used rectangular X-band cavity operating in the TE_{102} mode has an internal diameter of 11 mm whilst the cylindrical cavity operating in the TE_{011} mode has a sample access diameter of 20 mm (the subscripts in the TE notation designate the number of half-wavelengths along the cavity dimensions). The profiles of the $\boldsymbol{B}_1$ and $\boldsymbol{E}_1$ fields for the TE_{102} mode are illustrated in Fig. 3.5. The L-band (1 GHz) EPR cavity has a 23 mm access diameter that enables biomedical imaging of small animals, whereas in contrast the W-band resonator can only accommodate sample tubes of about 0.5 mm inner diameter.

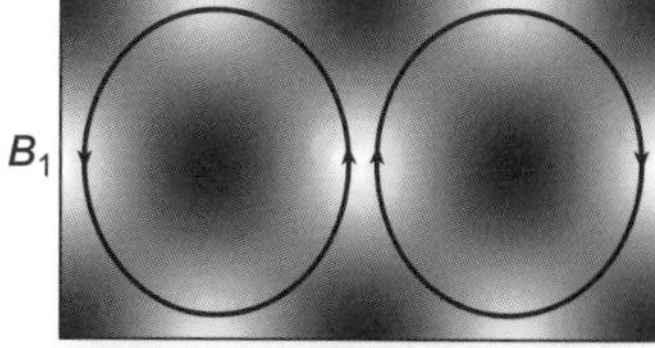

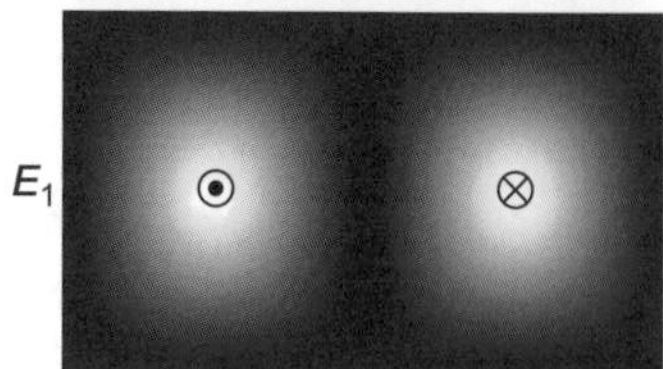

Fig. 3.5 Magnetic ($\boldsymbol{B}_1$) and electric ($\boldsymbol{E}_1$) field gradients for the TE_{102} mode. Lighter shading highlights regions of higher field densities.

3.3 Experimental settings

Microwave frequency

The choice of operating frequency will be determined by the nature of the sample and what information is sought from the resulting spectra. The most commonly used X-band frequency is ideally suited for measuring organic radicals in solution (which have g values close to free spin). However, there are several advantages to recording spectra at higher (and sometimes lower) frequencies, most notably the increased resolution of the anisotropic g values, as illustrated below in Fig. 3.6 for a system with similar g values (see Chapter 5). At X-band, the g values are virtually superimposed on each other, whereas at higher frequencies considerable resolution of the g values can be achieved. Additionally, high-frequency EPR has an increased absolute sensitivity, although the decrease in sample volume (due to smaller resonator dimensions) usually offsets any sensitivity enhancement gains. For systems containing multiple unpaired electrons, frequencies higher than W-band may be required.

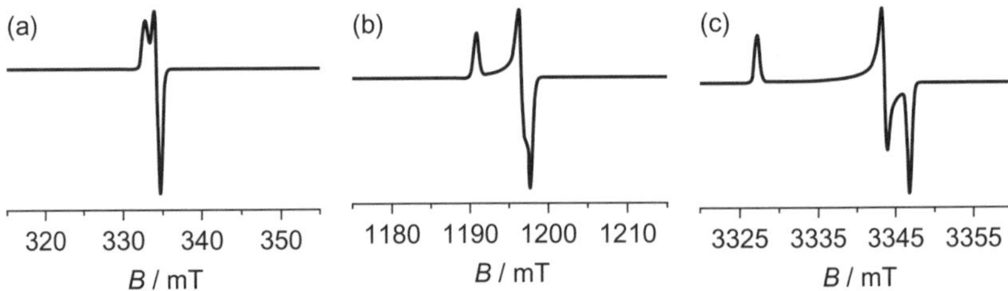

Fig. 3.6 Multi-frequency EPR spectra of a spin system characterized by similar g values, recorded at (a) X-band (9.5 GHz), (b) Q-band (34 GHz), and (c) W-band (95 GHz).

Microwave power

The microwave power incident upon the sample needs to be optimized in order to achieve the highest possible signal-to-noise (S/N) ratio, and to avoid possible lineshape distortions. At low power levels, the signal amplitude is related to the microwave power (P) by a $P^{1/2}$ function. Therefore the signal amplitude can be enhanced with increased microwave power (Fig. 3.7). However, at higher incident powers care should be taken to avoid *saturating* the spin system, whereby the populations of the electron Zeeman states become equal due to insufficient relaxation of the excited spins. Power saturation can lead to lineshape distortions and a decrease in signal intensity.

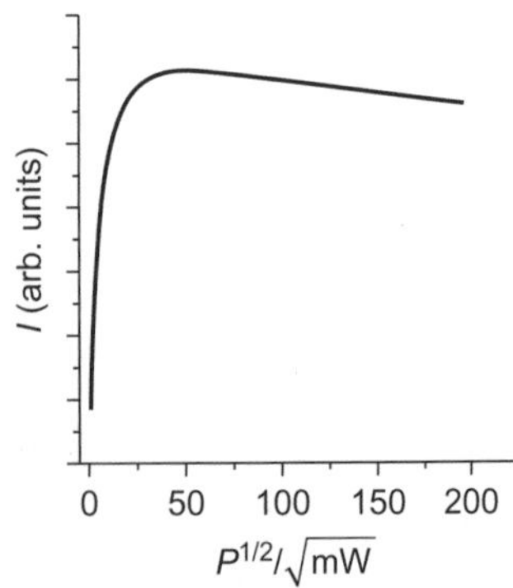

Fig. 3.7 Signal intensity of DPPH (see Chapter 4, Fig. 4.22 for structure) as a function of increasing microwave power, illustrating the $P^{1/2}$ dependency at low power and nearing saturation at high power.

Paramagnetic species with long T_1 times (e.g. organic radicals) may need to be recorded at low incident microwave powers (often <1 mW) to avoid saturation. By comparison, samples containing transition metal ions have much shorter T_1 times and may therefore be recorded using higher powers (often >10 mW). The different saturation characteristics allow the user to deconvolute the EPR spectra containing overlapping resonances from different paramagnetic centres using differential saturation. By recording the EPR spectrum of a sample at multiple incident powers, the saturation plots may allow assignment of the individual resonances to the different species.

Field modulation and modulation amplitude

CW EPR spectroscopy is performed using phase-sensitive detection to enhance the sensitivity. Two spectrometer parameters associated with the phase-sensitive detection have to be carefully selected for optimum signal amplitude and resolution, namely the *field modulation* and the *modulation amplitude*. A small additional oscillating magnetic field, referred to as the field modulation (B_m), is applied to the external magnetic field (B) typically at a frequency of 100 kHz (Fig. 3.8a). As the external field increases from B_{m1} to B_{m2}, the detector output increases from I_1 to I_2. The output also oscillates

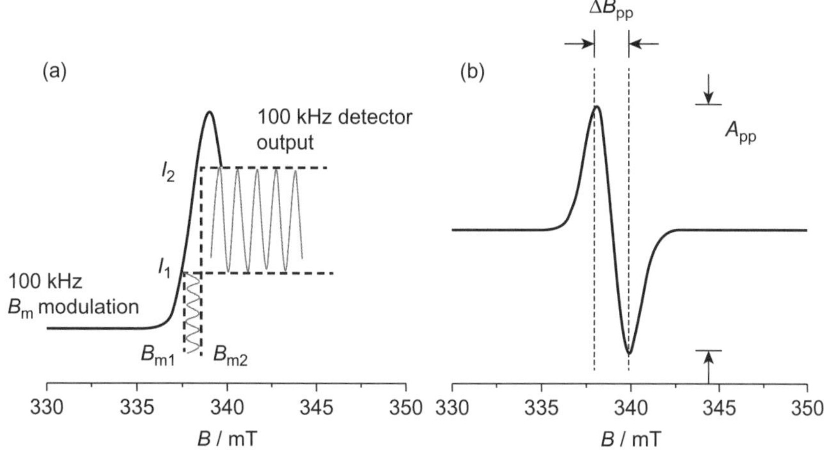

Fig. 3.8 (a) 100 kHz field modulation (B_m) is applied on the magnetic field B. (b) The peak-to-peak amplitude (A_{pp}) approximates the gradient of the absorption curve, producing a first derivative profile.

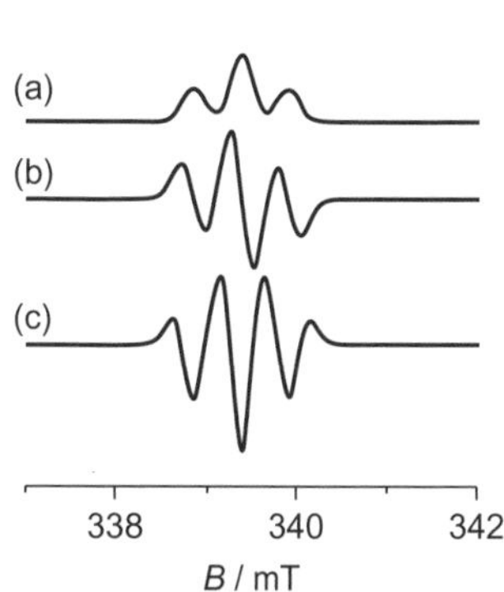

Fig. 3.9 Field modulation enables the EPR spectra to be recorded in (a) absorption, (b) first, or (c) second derivative modes.

at 100 kHz, yielding a peak-to-peak amplitude (A_{pp}) that approximates the gradient of the absorption curve (Fig. 3.8a), leading to a first derivative profile (Fig. 3.8b).

The signal channel, using phase-sensitive detection, only detects output signals with the same modulation (i.e. 100 kHz) as that of the reference arm signal (Fig. 3.2). Therefore signals arising from noise, principally of low frequency, are suppressed resulting in much greater S/N ratios. It is also possible to record EPR spectra in the second derivative mode (Fig. 3.9), which further enhances rapidly changing features in the spectrum and can be useful for resolving very narrow hyperfine lines. It should be noted that the field modulation is unique to CW spectroscopy. Spectra recorded on pulsed spectrometers are presented with absorption profiles.

Increasing the modulation amplitude, B_m, of the field is a useful method of increasing the peak-to-peak derivative line amplitude A_{pp} (i.e. it improves the S/N ratio). However, the user should be aware that the peak-to-peak linewidth, ΔB_{pp}, simultaneously increases with accompanying distortions to the lineshape, and eventually a decrease in the signal intensity (Fig. 3.10). Therefore, the optimum setting of B_m (0.01–2 mT) is a balance between maximizing the signal intensity whilst ensuring that narrow features are still resolved. If a modulation amplitude larger than the splitting between two EPR signals is used, then resolution of the two signals is lost. In this case, B_m should always be set to a value less than the width of the narrowest EPR line (ΔB_{pp}), or if accurate lineshape analysis is required $B_m < ½\Delta B_{pp}$. If the signal intensity is very weak, then B_m can be increased until a maximum A_{pp} is observed ($B_{m(max)}$). A value of $¼B_{m(max)}$ provides a suitable compromise between sensitivity and resolution. X-band spectrometers are typically limited to a modulation amplitude of 2 mT to avoid damaging the coils.

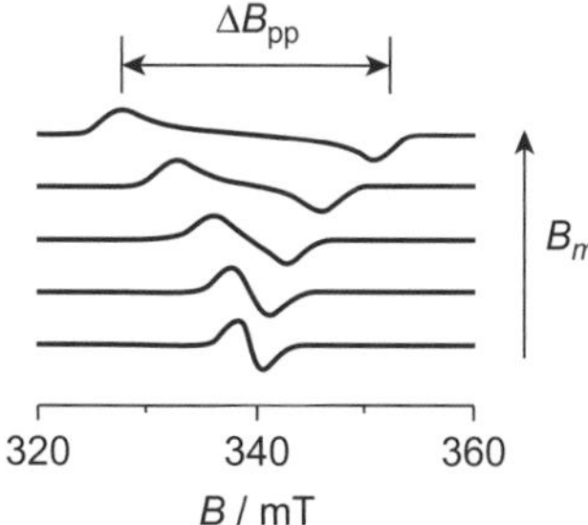

Fig. 3.10 Increasing the modulation amplitude B_m of the frequency modulated input leads to a simultaneous increase in ΔB_{pp}.

The modulation amplitude should be set to: $B_m < \Delta B_{pp}$ for resolution of narrow features, and $B_m < ½\Delta B_{pp}$ for accurate lineshape analysis.

Spectrometer receiver gain

The signal amplitude can be increased through adjustment of the spectrometer receiver gain. The receiver gain should be set so that the signal is maximized on the display, but care should be taken not to overload the amplifier, which is identified by a 'clipped' signal. In order to resolve all of the features in an EPR spectrum, including for example weak satellite lines or half-field transitions (see Chapter 7), it may be necessary to record several spectra at different receiver gain settings (one may also need to change the field sweep settings to focus on a particular transition or region).

Conversion time and time constant

The conversion time is the amount of time the digital-to-analogue converter spends accumulating a signal at each magnetic field position before it moves to the next field value. The number of data points acquired per spectrum was limited in older generation spectrometers to base two options (e.g. 256, 512, 1024, etc.), although modern instruments allow discrete values to be selected. In order to decrease the acquisition times, it is favourable to reduce the number of data points, but a larger number of points should be used for signals that have very narrow linewidths to ensure good resolution. Typically, in a CW measurement the full width of the spectrum is divided into 4096 data points, although as few as 512 points can accurately describe the lineshape in the EPR spectra which only contain broad peaks. A long conversion time is required when spectra contain weak signals. Conversely, a short conversion time is required for studies of transient species.

Sweep time = conversion time × number of data points.

The time constant acts to filter out noise. Therefore, increasing the time constant can lead to improved appearances, but a balance is required to ensure that the entire signal is not accidentally filtered and the lineshape is not distorted. Ideally, the time constant should be at least ten times smaller than the time required to scan through the narrowest EPR line.

A time constant: conversion time ratio of 1:4 should ensure that the spectrum is not distorted.

Signal intensity and spin concentration

It is often of interest to quantify the number of paramagnetic centres contributing to the observed EPR signal, as either relative or absolute values. However, the signal intensity is dependent on many factors, including the intrinsic characteristics of the sample (e.g. the anisotropy of **g**, **A**, or **D**, the relaxation rates, the power saturation properties) and also on various spectrometer settings (e.g. receiver gain, choice of microwave frequency, modulation amplitude). Therefore, the user should be aware of these different factors when performing quantitative studies. In practice, absolute concentrations are rarely reported, and it is often sufficient to make comparisons of signal intensities with respect to a suitable standard of known concentration.

Anisotropies of **g**, **A**, and **D** are dealt with in Chapters 5 and 7.

To minimize any potential sources of error, the standard should be recorded under identical operating conditions (i.e. using the same solvent, volume, temperature and power as the sample of unknown concentration) and placed in an identical position within the cavity. The double integral of the resulting EPR spectrum can then be determined, ensuring a sufficiently wide integral width is employed to account for broad Lorentzian linewidths (see Chapter 8) and low-amplitude hyperfine lines. Alternatively, if the EPR profile remains constant across the series of unknowns, the peak-to-peak amplitude, A_{pp}, can be compared between samples. The determination of the absolute number of spins giving rise to an EPR signal is even more complex than referencing to concentration standards, and requires the reference sample to have similar EPR properties to the unknown.

The sample concentration required for sufficient S/N ratios will vary with the physical state (e.g. liquid or frozen solution), the spectral width of the sample (e.g. systems with large **g**-anisotropy and/or complex hyperfine structure will require larger concentrations), and the operating frequency of the spectrometer. If the sample can be obtained in large quantities (which is often not the case for biological samples such as proteins or enzymes), it is advisable to perform experiments using high concentrations to minimize acquisition time. However, spin–spin interactions which lead to line broadening should be avoided; therefore it is recommended to use concentrations of approximately $<10^{-2}$ M (e.g. transition metals in the solid state) to 10^{-4} M (e.g. free radicals in solution).

For further details on obtaining accurate quantitative results from CW EPR experiments, the reader is referred to the text *Quantitative EPR* (see Bibliography). Suggested spectrometer settings for operation at X-band frequency are provided in Table 3.2.

Table 3.2 Suggested spectrometer settings for operation at X-band.

Settings	Organic radical	[a]Transition metal ion
Centre field	340 mT	300 mT
Sweep width	10 mT	500 mT
MW power	<5 mW	>10 mW
Mod frequency	100 kHz	100 kHz
Mod amplitude	0.1 mT	0.4 mT
Time constant	10 ms	10 ms
Conversion time	40 ms	40 ms
Sample		
Concentration	[b]10 μM [c]0.1 mM	[b]0.5 mM [c]0.25 mM
Volume	100 μL	100 μL
Temperature	100–400 K	4–298 K

[a]Settings vary considerably more for transition metal ions than organic radicals; [b]liquid phase; [c]frozen solution.

3.4 Sample preparation

Physical state of sample

One of the significant advantages of EPR spectroscopy is that samples can be recorded as solids (including single crystals and amorphous powders), liquids, frozen solutions and even gases. Gas-phase EPR spectra of S-state atoms tend to result in narrow EPR lines due to the relatively long relaxation times T_1 (see Chapter 2). As intermolecular electron-electron spin interactions can affect the profile of the EPR spectra (see Chapter 7), measurements are usually performed on samples that are magnetically dilute. For insoluble solid materials this may require doping the sample of interest into a diamagnetic isostructural host. For soluble materials, an easy method of achieving magnetic dilution is to dissolve the sample into a suitable solvent such that the solvent molecules act to separate the paramagnetic centres.

The choice of solvent is often determined by the sample solubility, but it is also important to consider the solvent dielectric properties (Table 3.3). For example, solvents of high dielectric constant can lead to difficulties in tuning the resonator due to unwanted absorption of the microwave energy, leading to dielectric heating loss. These samples should be recorded using an EPR *flat-cell*, which is designed to achieve maximum interaction with the $\boldsymbol{B}_1$ component and minimum infringement on the $\boldsymbol{E}_1$ component of the microwaves. The dielectric loss of solvents decreases with decreasing temperature, therefore the tuning characteristics may improve with a reduction in operating temperature. To prevent the formation of small pockets of crystalline phases that can lead to shortened relaxation times and distortions or artefacts in the signal, it is common practice to record samples in mixed-solvent systems such as toluene: dichloromethane (1:1 ratio), methanol:dimethyl sulfoxide (1:1), dichloromethane: acetonitrile (1:1), water:glycerol (between 4:1 and 1:4), water:propylene glycol (1:1), diethylether:ethanol (3:1), and ethanol:methanol (ratios of 4:1, 5:2, 1:9).

Table 3.3 Physical properties of common solvents.

Solvent	m.p./K	Dielectric constant	Viscosity/ cP
Acetone	178.3	20.5	0.324
Dichloro-methane	176.5	9.08	0.413
Toluene	178.1	2.4	0.585
Methanol	175.3	32.6	0.594
Water	273.0	78.8	1.002
Ethanol	158.9	24.3	1.197
Glycerol	291.0	45.5	1490

Temperature

It is often necessary to record the EPR spectrum at variable temperatures. In the first instance, decreasing the sample temperature (T_s) will increase the sensitivity through an increase in the electron spin polarization, as described by the Boltzmann relation (see Chapter 2). However, the motivation for performing low temperature EPR spectroscopy may also be driven by other factors. For example, it may be necessary to cool the sample in order to slow down the relaxation rates (see section 2.5), to enable detection of thermally accessible triplet states (see Chapter 7), or to determine the anisotropic **g** and **A** parameters from a frozen solution (see Chapters 5 and 6). Low

Sealed Pasteur pipettes or glass or quartz capillaries are alternatives to expensive EPR *flat-cells* for X-band and higher frequency measurements.

temperatures can be achieved either using liquid nitrogen (77 K) or liquid helium (4 K) and are controlled through a variable temperature control unit. To reduce the dependence on rapidly declining (and expensive) liquid helium reserves, cryogen-free systems that utilize gaseous helium as the heat transfer medium are now commercially available.

3.5 Summary

- The main components of a CW EPR spectrometer are the microwave source, the magnet, and the microwave resonator.
- CW EPR spectroscopy employs phase-sensitive detection, so the resulting spectra are usually recorded as first derivatives.
- The choice of operating frequency will be sample dependent.
- Several inter-related instrument settings need to be optimized in order to maximize the quality of the EPR signal.
- Signal intensities can be increased through recording spectra at low temperatures.
- Quantitative EPR requires a thorough understanding of the spectrometer optimization parameters and the nature of the spin system itself.

3.6 Exercises

3.1) Using the resonance equation (eqn 2.9), calculate the g and B values in the following:

   a) $\nu = 9.486$ GHz, $B = 0.3345$ T, $g = ?$

   b) $\nu = 34.86$ GHz, $g = 2.0023$, $B = ?$

3.2) The centre of the EPR spectrum of atomic hydrogen lies at 329.12 mT in a spectrometer operating at 9.2231 GHz. What is the g value of the atom?

3.3) The benzene radical anion has a g value of 2.0025. At what field should you search for resonance in a spectrometer operating at:

   a) 9.302 GHz,

   b) 33.67 GHz?

3.4) At thermal equilibrium (and under the influence of an external magnetic field) the relative difference in population between the two Zeeman energy levels is given by the Maxwell–Boltzman law. Calculate the population ratio between the two levels at the two operating temperatures 298 K and 4 K, for a paramagnetic species at 300 mT and 1300 mT. Comment on the significance of the values for the intensity of the EPR signal.

3.5) For an unpaired electron ($g = 2.25$) interacting with a proton ($g_N = 5.58$), explain the relative magnitude of the EZ and NZ interactions.

Isotropic EPR spectra of organic radicals

4.1 Introduction

Chapter 2 showed that an unpaired electron in a magnetic field experiences a hyperfine interaction with the nearby nuclei possessing a nuclear spin. This interaction results in splitting of the EPR spectra into several lines (e.g. Chapter 2, Fig. 2.5), in analogy with the *J* coupling in NMR.

For organic radicals in particular, the g value is usually close to that of a free electron ($g_e = 2.0023$). Therefore all organic radicals give EPR signals at roughly the same position in the EPR spectrum and even accurate measurements of the g factor provide limited structural information. The splitting originating from the hyperfine interaction thus provides the main tool for assigning the structure of organic radicals.

Chapter 2 described two contributions to the hyperfine interaction: the Fermi contact (isotropic interaction originating from the non-zero electron density at nucleus) and dipole–dipole (anisotropic interaction between magnetic dipole moments of the electron and the nucleus). The anisotropic interaction averages out to zero for rapidly tumbling radicals, e.g. for small organic radicals in low viscosity solvents. In describing EPR spectra of organic radicals in this chapter, we will therefore focus on the Fermi contact (isotropic) hyperfine interaction.

After introducing the effect of a hyperfine interaction on the EPR spectra, this chapter explains how to build *splitting diagrams*, measure hyperfine constant values and simulate the spectra, and what structural information can be obtained from these data. The second part of the chapter will discuss strategies for the generation and detection of short lived organic radicals.

4.2 Isotropic hyperfine interaction

Consider an unpaired electron that experiences a hyperfine interaction with a nucleus of spin I. As seen in Chapter 2, this system has $2I + 1$ energy levels for each m_S value ($m_S = +½$ and $m_S = -½$). There are $2I + 1$ EPR transitions between these levels (remembering that EPR transitions are those that flip the electron spin, i.e. $\Delta m_S = \pm 1$) and therefore *the EPR spectrum of a radical that experiences a hyperfine interaction with a nucleus of spin I shows 2I + 1 lines.*

For commonly used microwave frequencies (e.g. X-band), the energy of the hyperfine interaction for organic radicals is much smaller than the energy of the electron

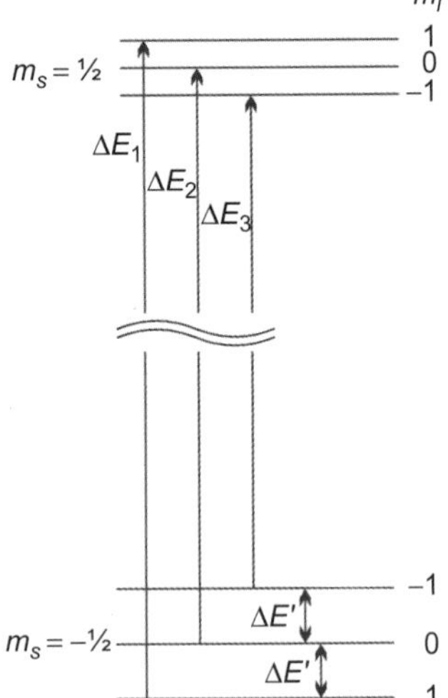

Fig. 4.1 In the high field approximation, the energy difference between EPR transitions (e.g. $\Delta E_1 - \Delta E_2$) is very small compared to the electron Zeeman interaction energy (e.g. ΔE_2) as illustrated for the two-spin system $S = ½$, $I = 1$. The relative position of the m_I levels assumes that the hyperfine constant is positive and the hyperfine term is larger than the nuclear Zeeman term (see Chapter 2).

Zeeman interaction (do not confuse electron and nuclear Zeeman inteactions, see Chapter 2). This is known as the *high field approximation*. The EPR transitions are thus characterized by very similar energy gaps (labelled ΔE_1, ΔE_2, and ΔE_3 in Fig. 4.1), and hence the probabilities of these transitions are very similar. The hyperfine lines in the EPR spectra therefore have equal intensity.

In addition, the m_I levels are evenly spaced (i.e. the energy gaps $\Delta E'$ in Fig. 4.1 are almost the same in the high field approximation). Therefore the distances between the $2I + 1$ lines (of equal intensity) in the EPR spectrum are equivalent. Examples of hyperfine splitting of EPR spectra are shown in Fig. 4.2.

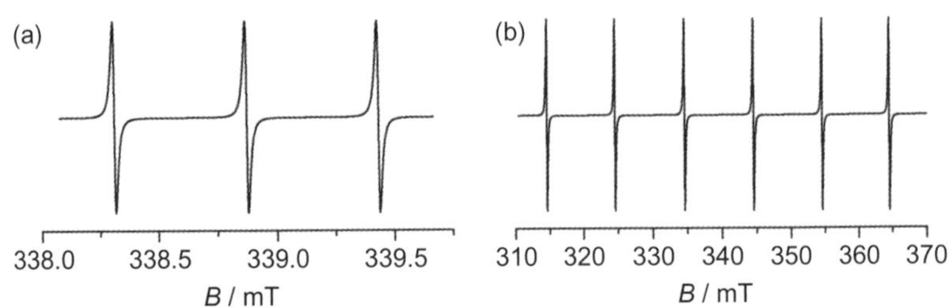

Fig. 4.2 EPR spectra of: (a) a nitrogen atom ($I = 1$) trapped inside a C_{60} molecule. The spectrum consists of three equidistant lines ($2I + 1 = 2 \times 1 + 1 = 3$) of equal intensity; and (b) Mn^{2+} ($I = 5/2$) doped MgO sometimes used as an EPR reference marker. The spectrum consists of six lines ($2I + 1 = 2 \times 5/2 + 1 = 6$) of equal intensity.

Splitting diagrams

EPR spectra can become more complex in the presence of several nuclei possessing nuclear spin $I \neq 0$. Fortunately, their interpretation is facilitated by the use of EPR splitting diagrams, which show how a signal is split into several lines due to a hyperfine interaction (in analogy with the diagrams commonly used in NMR spectroscopy to indicate J coupling). Splitting diagrams can be thought of as simplified energy level diagrams (such as Chapter 2, Fig. 2.5) rotated by 90°. Let us build a splitting diagram for a radical with the unpaired electron coupled to two nuclei (A and B) with nuclear spins $I_A = ½$ and $I_B = 1$. An example of such a system would be a radical anion of monodeuterated methanal $[CHD{=}O]^{\bullet-}$ ($I_H = ½$ and $I_D = 1$). The EPR signal is split by the $I_A = ½$ nucleus into $2I + 1 = 2$ lines, and so a splitting diagram shows a doublet. Each line of this doublet is further split by the $I_B = 1$ nucleus into three lines of equal intensity, thus giving a doublet of triplets. The complete splitting diagram is shown above the spectrum in Fig. 4.3.

Building a splitting diagram for a radical interacting with two nuclei, $I_A = ½$ and $I_B = 1$: the line is split first into a doublet due to $I_A = ½$ nucleus and then into a triplet due to $I_B = 1$ nucleus (diagram above the spectrum)

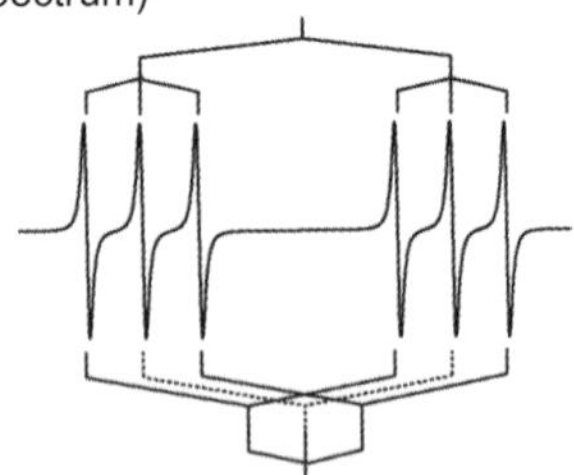

Alternative way to build the diagram: first split the line into a triplet due to $I_B = 1$, then each component of this triplet into a doublet due to $I_A = ½$. The central doublet is shown with a dotted line for clarity (diagram below the spectrum)

Fig. 4.3 Two ways of building splitting diagrams for $I_A = ½$ and $I_B = 1$ systems.

An alternative splitting diagram for the same system is shown below the spectrum in Fig. 4.3. Here, the interaction of the unpaired electron with the $I_B = 1$ nucleus is considered first giving three lines of equal intensity, with each line being further split into a doublet by the $I_A = ½$ nucleus. Although the two splitting diagrams look different, they lead to identical peak positions with identical EPR spectra and are both correct. It is usually easiest to start the diagram with the largest splitting (which depends on the magnitude of the hyperfine interaction; in Fig. 4.3 $a(A) > a(B)$).

Sometimes resonance lines overlap which can make the assignment of EPR spectra more difficult. For example, let us build a splitting diagram for the substituted nitroxide radical shown in Fig. 4.4(a). The nitrogen and two inequivalent protons should split the

EPR signal into a triplet ($I(^{14}N) = 1$) and two doublets, ($I(^{1}H) = ½$). Therefore, one would expect to see a triplet of doublet of doublets resulting in a total of $3 \times 2 \times 2 = 12$ lines. However the actual spectrum only shows 10 lines, as two pairs of lines overlap giving two signals of higher intensity. The spectrum and the splitting diagram are shown in Fig. 4.4(b). The apparent overlapping of lines is partially due to the large line width in Fig. 4.4(b) (e.g. the peaks are significantly broader than in Fig. 4.2). The factors that determine line widths in EPR spectroscopy are quite complex and are considered in more detail in Chapter 8.

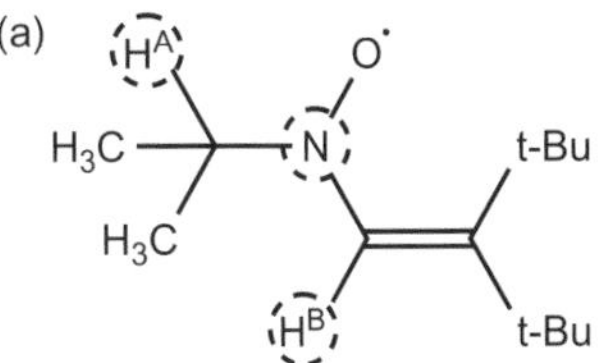

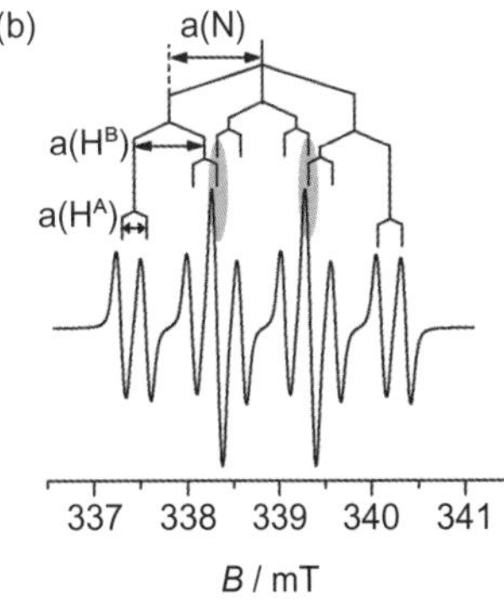

Fig. 4.4 (a) Substituted nitroxide radical (nitroxide group is >N–O˙). The nuclei participating in the hyperfine interaction observed by EPR are circled. (b) EPR spectrum and splitting diagram. The overlapping lines are highlighted with grey ovals.

After H. G. Aurich, K. Hahn, and K. Stork, *Angew. Chem. Int. Ed. Engl.* 1975, **14**, 551.

Hyperfine interaction with several equivalent nuclei

The hyperfine interaction with several equivalent nuclei (i.e. nuclei in identical chemical environments) is considered in the same way as the interaction with non-equivalent nuclei described above. Let us build a splitting diagram for the hyperfine interaction of the unpaired electron with three equivalent $I = 1$ nuclei (which would for instance be observed for the $^{\cdot}CD_3$ radical) (Fig. 4.5). As the interaction with each $I = 1$ nucleus is independent from the other interactions, we will consider each interaction in turn. The first $I = 1$ nucleus will split the signal into three lines of equal intensities, i.e. a triplet. Each of these lines will be further split into a triplet by the interaction with the second nucleus. As the two nuclei are equivalent, the distances between the components of both triplets will be the same. Hence the lines in the middle of the multiplet overlap giving rise to lines of double and triple intensity (i.e. the combined intensity of the overlapped lines). Therefore, the spectrum of a radical with two equivalent $I = 1$ nuclei would give five lines in a 1:2:3:2:1 ratio (Fig. 4.5). Each of these lines will be split into further triplets by the interaction with the third $I = 1$ nucleus. Again, the lines in the middle will overlap ultimately giving rise to a 1:3:6:7:6:3:1 septet. In general, *n equivalent nuclei with spin I will split EPR signals into 2nI + 1 lines*. In the special case of coupling with several equivalent $I = ½$ nuclei (such as ^{1}H), the number of lines and their relative intensity is given by the binomial expansion coefficients, which are conveniently visualized by Pascal's triangle (Fig. 4.6).

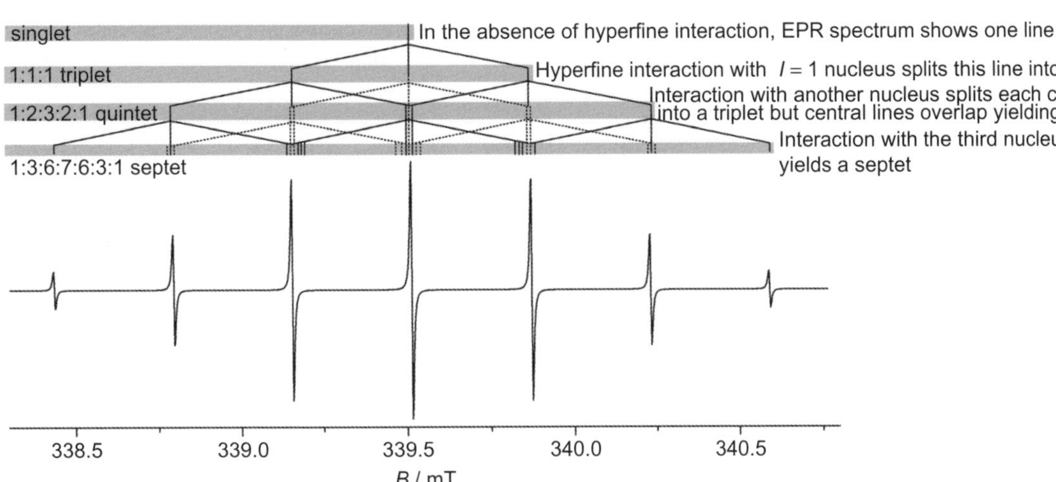

Fig. 4.5 Splitting diagram and EPR spectrum for a radical with three equivalent $I = 1$ nuclei, such as $CD_3^{\cdot}$.

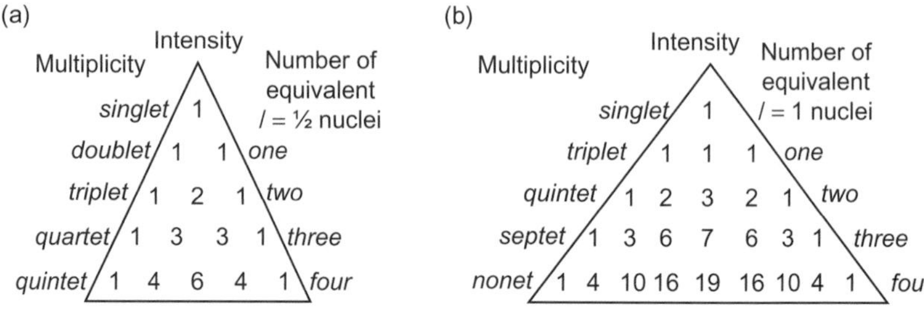

Fig. 4.6 Pascal's triangle predicts the intensity and multiplicity for (a) radicals with several equivalent $I = ½$ nuclei and (b) a similar diagram for $I = 1$ nuclei.

Equations for converting hyperfine constants a into different units are given in Appendix A.

Understanding how splitting diagrams are built is essential for the assignment of EPR spectra and takes some practice. The reader is directed to the exercises section and online resources such as www.EPRsimulator.org.

Hyperfine coupling constants

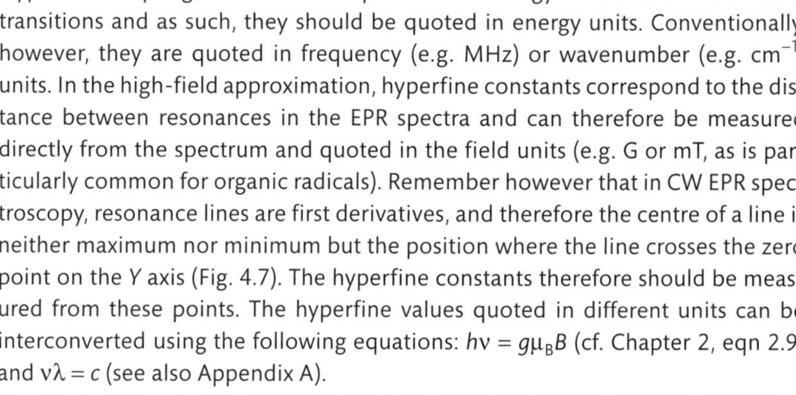

Hyperfine coupling constants correspond to the energy differences between EPR transitions and as such, they should be quoted in energy units. Conventionally, however, they are quoted in frequency (e.g. MHz) or wavenumber (e.g. cm^{-1}) units. In the high-field approximation, hyperfine constants correspond to the distance between resonances in the EPR spectra and can therefore be measured directly from the spectrum and quoted in the field units (e.g. G or mT, as is particularly common for organic radicals). Remember however that in CW EPR spectroscopy, resonance lines are first derivatives, and therefore the centre of a line is neither maximum nor minimum but the position where the line crosses the zero point on the Y axis (Fig. 4.7). The hyperfine constants therefore should be measured from these points. The hyperfine values quoted in different units can be interconverted using the following equations: $h\nu = g\mu_B B$ (cf. Chapter 2, eqn 2.9) and $\nu\lambda = c$ (see also Appendix A).

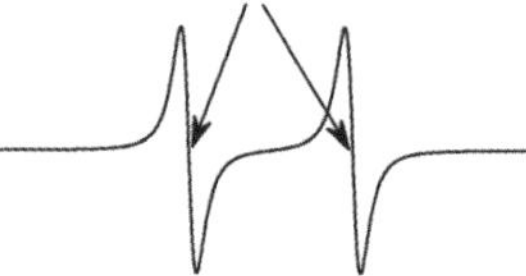

Fig. 4.7 Hyperfine constants can be measured directly from the EPR spectra (see Chapter 2).

The factors affecting the strength of the hyperfine interaction are discussed in section 4.5. However, in most cases this interaction is not visible for nuclei further than two bonds away from the atom with the unpaired electron. For example, in the EPR spectrum of the n-propyl radical $CH_3CH_2CH_2^{\bullet}$, one would only expect to see splitting from the protons of the two CH_2 groups. These protons are termed α- and β-protons, counting from the atom with the unpaired electron. The hyperfine interaction of the unpaired electron with the γ-protons (i.e. CH_3 protons) which are three bonds away from the radical centre will be expected to give a negligible hyperfine coupling constant. The EPR spectrum of the propyl radical therefore shows a triplet of triplets (Fig. 4.8).

Very weak hyperfine constants can be determined using advanced EPR techniques, see Chapter 9.

Hydrocarbon radical nomenclature: counting starts with the atom with unpaired electron, for instance $CH^{\gamma}_3CH^{\beta}_2CH^{\alpha}_2{}^{\bullet}$.

It is interesting to note that the β-hyperfines in the propyl radical spectrum are larger than the α-hyperfines. This is explained in section 4.5.

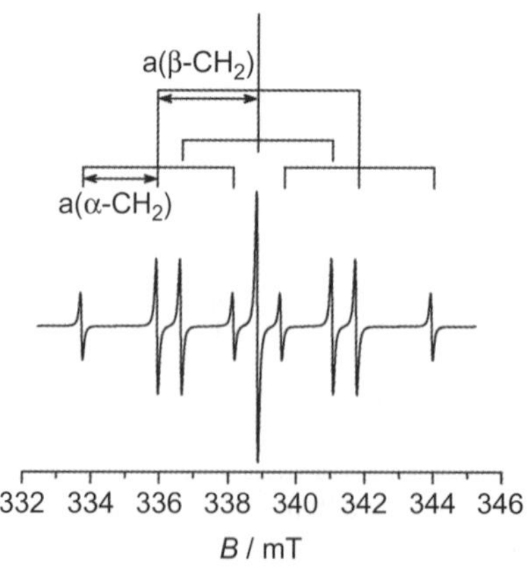

Fig. 4.8 EPR spectrum and splitting diagram for the $CH_3CH_2CH_2^{\bullet}$ radical.

4.3 Localized and delocalized radicals

The unpaired electron in a radical occupies the SOMO (singly occupied molecular orbital). In many radicals this orbital is largely localized on a particular atom (e.g. in a p-orbital or a hybridized orbital). In other words, the unpaired electron density (i.e. the spin density) is about unity on one atom and almost zero on all other atoms. Alkyl radicals such as $CH_3CH_2^{\cdot}$ (where the electron density is localized on the methylene carbon atom) exemplify this type of radical (Fig. 4.9a). In other radicals, however, the SOMO can be delocalized over the whole molecule. The benzyl radical ($PhCH_2^{\cdot}$) is an example of a delocalized radical, where due to resonance delocalization, there is different spin density on each of the carbon atoms (Fig. 4.9b). All protons in the benzyl radical can thus be considered α with respect to the unpaired electron, as they are all attached to a carbon atom that contains some unpaired electron density.

(a) (b)

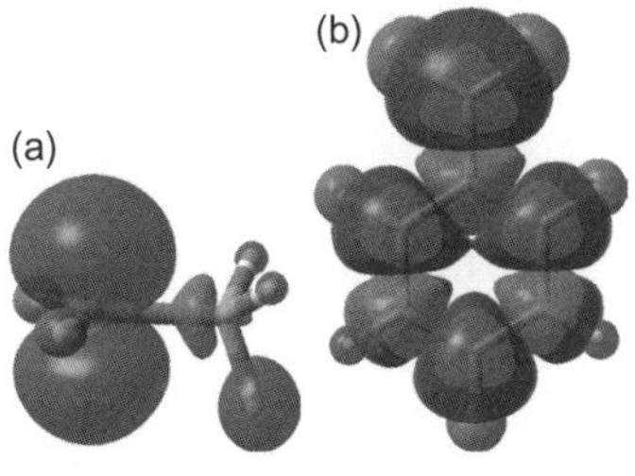

Fig. 4.9 Distribution of spin density in (a) ethyl radical (localized) and (b) benzyl radical (delocalized).

In delocalized radicals, a hyperfine interaction will be observed for nuclei with $I \neq 0$ attached to atoms in the α- and β-position with respect to the unpaired electron. In the benzyl radical (Fig. 4.10), the two ortho-protons are attached to carbons with the same spin density and their chemical environment is identical; hence they are equivalent and will split the EPR signal into a 1:2:1 triplet. Meta-protons are also equivalent and yield triplet splitting. The fast rotation around the C–C bond linking the aromatic ring to the methylene group makes the two protons H_α equivalent; hence a triplet splitting will also arise from a hyperfine interaction with this group. The single para-proton will split the signal into a doublet. The EPR spectrum of the benzyl radical should thus give a triplet (H_α) of triplets (H_{ortho}) of triplets (H_{meta}) of doublets (H_{para}) resulting in $3 \times 3 \times 3 \times 2 = 54$ lines (section 4.4).

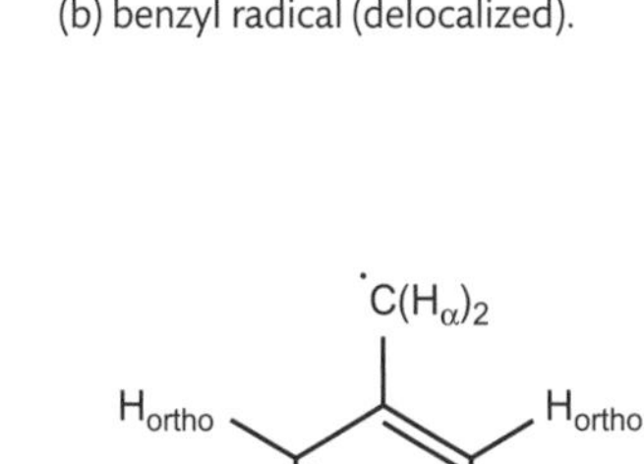

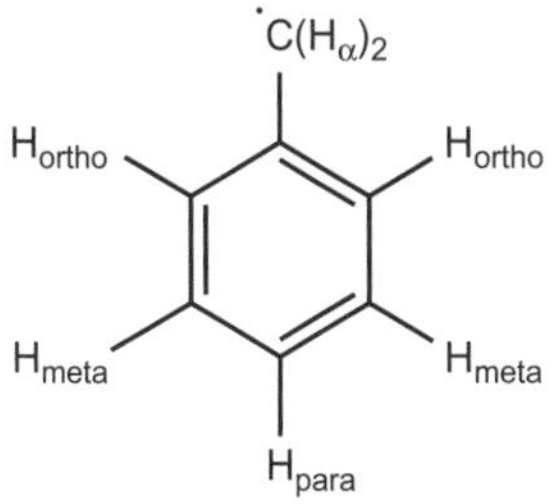

Fig. 4.10 The structure of the benzyl radical.

4.4 Assignment of complex EPR spectra

The EPR spectra of organic radicals, in particular of delocalized radicals, can be very complex and include hundreds or even thousands of lines. Often careful analysis of these spectra makes it possible to achieve complete assignment. The routine for analysing and interpreting an isotropic EPR spectrum arising from an organic radical is discussed below.

Routine for analysing isotropic EPR spectra

1. *If the radical structure is known*, identify all spin-active ($I \neq 0$) nuclei in the α- and β-position. Count the number of equivalent spin-active nuclei in each environment and use the $2nI + 1$ rule and Pascal's triangle to determine the multiplicity and relative intensity of the hyperfine lines for each environment.
2. If the overall spectrum is not symmetrical with respect to the spectral centre, it probably contains signals from several radicals which need to be analysed separately.
3. Measure the separation between the first two lines at low field in the spectrum. This distance corresponds to the smallest hyperfine coupling (labelled multiplet A).
4. Check if more lines are located at the same distance (i.e. hyperfine coupling) upfield from the second line identified in step (3); they are likely to belong to the same multiplet A.
5. Check the intensities of the lines in the multiplet A. If the multiplet is not symmetrical, it probably overlaps some lines from another multiplet.

(continued...)

Never start interpreting the spectrum by examining the middle lines; always start at the low or high field extremes of the spectrum.

All lines in the same hyperfine pattern are equally spaced in the high-field approximation.

Lines often overlap in EPR spectra.

6. Measure the distance from the outermost line to the first line which does not form part of multiplet A. This distance corresponds to the hyperfine coupling constant of the next multiplet (labelled B).

All hyperfine multiplets are centrosymmetric.

7. The hyperfine pattern of multiplet A will be superimposed on each component of multiplet B.
8. Measure hyperfine coupling B towards higher field until all components of multiplet B have been identified.

Repeat steps 6–9 until all multiplets have been identified.

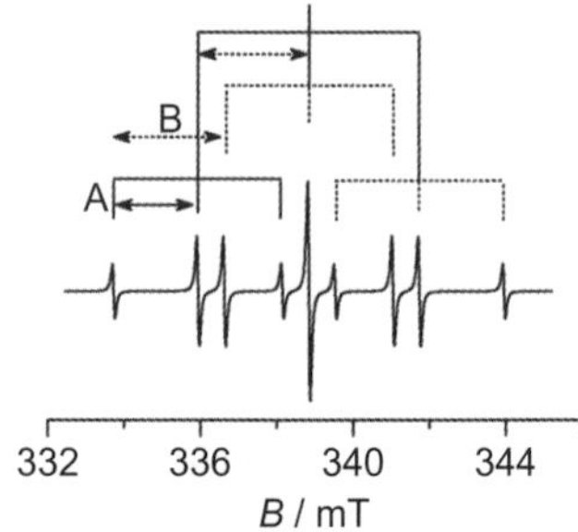

Fig. 4.11 Assigning the EPR spectrum of the $CH_3CH_2CH_2^{\cdot}$ radical.

After R. W. Fessenden and R. H. Schuler, *J. Chem. Phys.* 1963, **39**, 2147.

The application of these rules can be illustrated with the *n*-propyl radical spectrum (Fig. 4.8). Two α- and two β-protons will each produce triplet splittings; hence the overall spectrum will consist of a triplet of triplets, e.g. $3 \times 3 = 9$ lines. The distance between the first and the second lines (Fig. 4.11, solid arrow) indicates the smallest hyperfine constant. A third line at the same distance at a higher field completes the first 1:2:1 triplet (multiplet A). The distance between the outermost line and the first line which does not belong to multiplet A (e.g. the third line) is hyperfine B (Fig. 4.11, lower dotted arrow). As multiplets A superimpose on each component of multiplet B, multiplets B connect the centres of multiplets A. In order to identify the remaining multiplets A (shown with dotted lines), we therefore measure out hyperfine B from the centre of the first multiplet A (Fig. 4.11, upper dotted arrow).

Let us now assign a more complex EPR spectrum of the benzyl radical (Fig. 4.12). A triplet of triplets of triplets of doublets is expected (see section 4.3). Start at the left edge of the spectrum. Three equidistant lines are easily recognized as triplet A with a 1:2:1 intensity ratio (Fig. 4.13a). Determine the distance between the outermost line and the first line that is not part of triplet A (Fig. 4.13b, solid arrow). Measure this distance from the centre of multiplet A (Fig. 4.13b, dashed arrow) and find the second triplet. Measure the same distance again to find the third triplet (Fig. 4.13c). We have identified the first triplet (B) of triplets (A) pattern based on a 1:2:1 intensity pattern.

Repeat the same procedure again. Determine the distance between the outermost line and the first line that is not part of the current pattern (triplet B or triplets A, Fig. 4.13d, solid arrow). Measure this distance from the centre of multiplet B (Fig. 4.13d, dashed arrow) and find the second triplet of triplets. It is shown with dotted lines in Fig. 4.13(d) for clarity. The two triplets of triplets in Fig. 4.13(d) have the

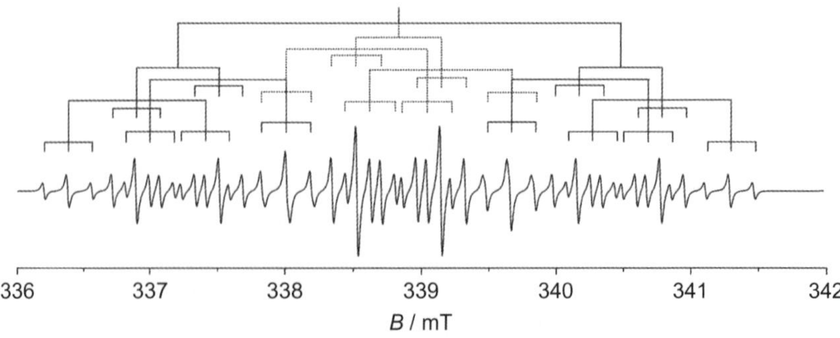

Fig. 4.12 The assigned EPR spectrum of the benzyl radical.

After P. J. Krusic and J. K. Kochi, *J. Am. Chem. Soc.* 1968, **90**, 7155.

same intensity; therefore they form a 1:1 doublet. We now have a doublet (C) of triplets (B) of triplets (A).

Repeating the same procedure again is more difficult, as some lines overlap. Therefore let us think back to our original expectation of a triplet of triplets of triplets of doublets. By following our assignment procedure we have so far found a doublet of triplets of triplets. There is one triplet splitting remaining. Therefore the peaks identified in Fig. 4.13(d) must be repeated in two other parts of the spectrum. Remembering that the spectrum is symmetric, we can draw an identical doublet of triplets of triplets in the high field part of the spectrum, and then find the remaining doublet of triplets of triplets (Fig. 4.12, dotted line) to complete the assignment. Two of the triplets shown with dotted lines around 338 mT and 339.7 mT overlap the solid line triplets and hence the total number of lines in the spectrum is 48 rather than 54. The hyperfine constants can now be measured from the distances between the lines and the multiplets.

(a)

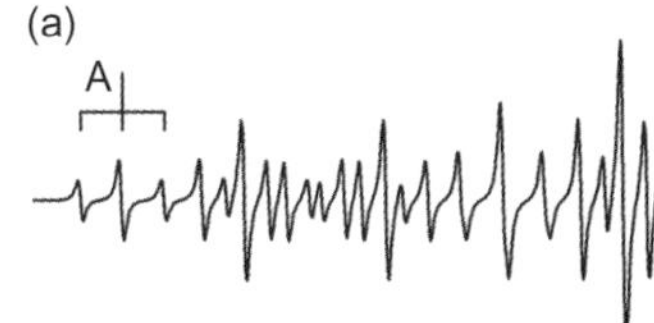

(b)

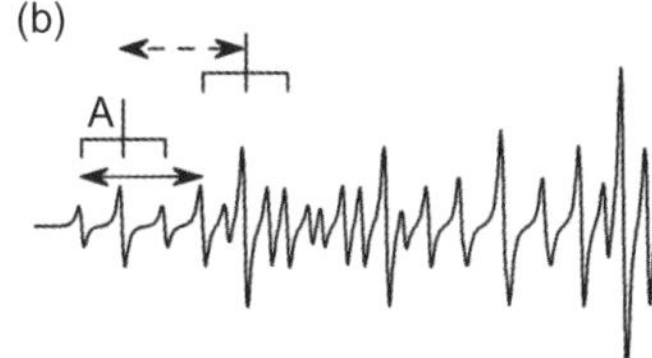

(c)

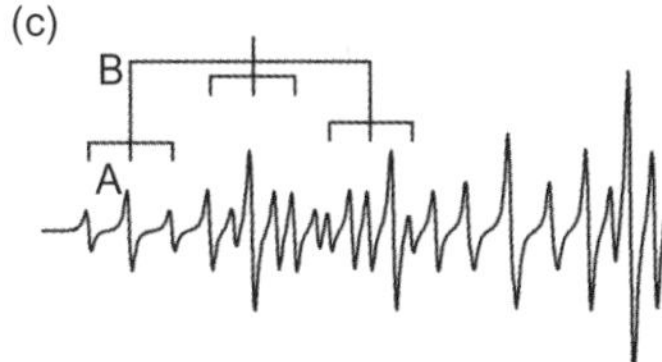

(d)

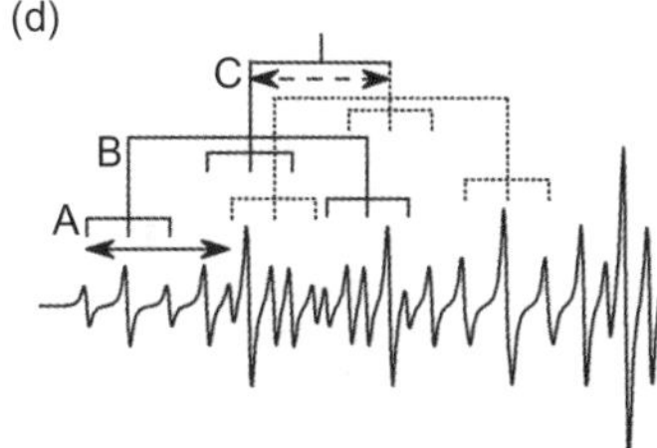

Fig. 4.13 Step-by-step assignment guide of the EPR spectrum of the benzyl radical.

Common problems with the analysis of complex EPR spectra

- Weak lines at the edge of the spectrum could be lost in the noise or even not recorded if the operator did not see them.
- The relative intensities in the multiplets can deviate from predicted values due to bad baseline or linewidth effects, discussed in Chapter 8.
- Line overlaps are commonly encountered in complex spectra.
- The high-field approximation (which leads to equidistant lines in the multiplets) is only valid for relatively small hyperfine constants.

Spectra simulations

Simulation of complex isotropic EPR spectra is an essential part of the spectrum assignment. Computer simulation makes it possible to accurately determine hyperfine constants, check the relative intensities of lines in multiplets, find any unassigned lines, and deconvolute multicomponent spectra into individual components.

There are a number of EPR simulation packages freely available on the internet.

WinSim (http://www.niehs.nih.gov/research/resources/software/tox-pharm/tools/) is a tool developed for Windows 3.1 but it is compatible with the newer versions of Windows. WinSim is useful for simulation of complex isotropic EPR spectra and fitting to experimental spectra.

EPR simulator (http://EPRsimulator.org) is an online script which works in any browser on any operating system. It allows the user to simulate EPR spectra and includes many features aimed at teaching the user to interpret EPR spectra through examples and problems.

Some commercial spectrometers include simulation tools as part of the spectrometer software.

Software accompanying the book *Biomolecular EPR Spectroscopy* by W. R. Hagen (http://www.bt.tudelft.nl/biomolecularEPRspectroscopy) includes a number of small programs designed for EPR beginners.

EasySpin (http://www.EasySpin.org) is an advanced simulation program enabling the user to simulate almost any kind of EPR spectra, including more advanced EPR techniques. EasySpin is a Matlab toolbox and requires the user to have Matlab licence.

Spectrum simulation is not a shortcut to spectrum assignment. It may be tempting to let the software do the assignment but this very rarely works; manual spectrum analysis and hyperfine constant measurement usually has to be done before any automatic spectrum fitting is attempted. The site www.EPRsimulator.org has a number of exercises that will help the reader to learn the art of assignment and simulation of EPR spectra of organic radicals.

4.5 Obtaining structural information from the hyperfine constants

In order to interpret the values of the hyperfine constants measured from the EPR spectrum, the mechanisms of the hyperfine interaction need to be considered in more detail compared to that presented in Chapter 2, section 2.4.

Fermi contact interaction

The Fermi contact describes the interaction between a nucleus and an unpaired electron that directly overlap (see section 2.4). This is only true for electrons in orbitals with some s-character. For example, the unpaired electron in a hydrogen atom is in the 1s-orbital that has some electron density at the proton, and so the electron spin can interact directly with the nuclear spin. This interaction results in a large hyperfine constant for the hydrogen atom. Similar interactions are observed in radicals with unpaired electron density in hybridized orbitals with s-character, such as sp^2 or sp^3 orbitals. For example, the unpaired electron in the ^{13}C-labelled formyl radical occupies an sp^2 orbital and thus the hyperfine interaction with the ^{13}C nucleus is explained by direct Fermi contact interaction. Radicals with the unpaired electron in an orbital with s-character are called σ-radicals.

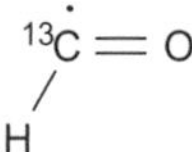

^{13}C-labelled formyl radical

In the Fermi contact interaction, the value of the hyperfine interaction is proportional to the spin density at the nucleus and the g_N factor of the nucleus (see eqn 2.20). This makes it easy to predict the values of hyperfine constants for isotopically-substituted radicals: as the distribution of electron density is the same for isotopomers, the ratio of hyperfine constants equals the ratio of the g_N factors. For instance, the EPR spectrum of a hydrogen atom trapped in synthetic quartz has an experimental hyperfine constant measured from the spectrum of $a = 52.06$ mT, whereas that of a deuterium atom has $a = 7.97$ mT (Fig. 4.14), in good agreement with the ratio of nuclear g factors ($g_N(\text{H}) = 5.5857$, $g_N(\text{D}) = 0.8574$).

Instead of nuclear g factors, gyromagnetic ratios can be used to predict the hyperfine constants for isotopically-substituted radicals. The *nuclear gyromagnetic ratio* γ_N (cf. Chapter 2, section 2.2) is an intrinsic property of a nucleus, $\gamma_N = g_N\mu_N/\hbar$. See *Nuclear Magnetic Resonance* primer for more details.

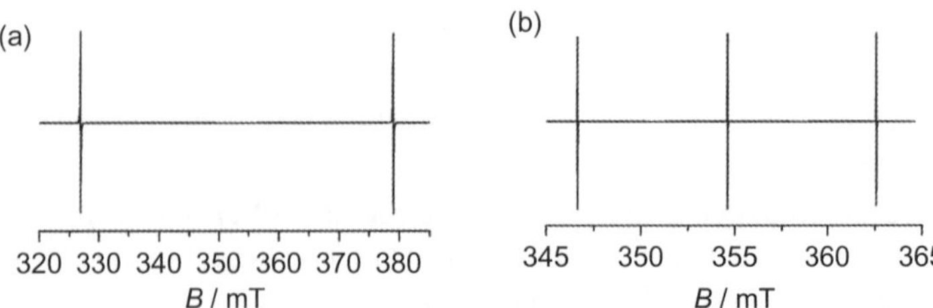

Fig. 4.14 EPR spectra of: (a) a hydrogen atom ($I = ½$), and (b) a deuterium atom ($I = 1$) in irradiated quartz with hyperfine constants of 52.06 mT (a) and 7.97 mT (b).

After J. Isoya, J. A. Weil, and P. H. Davis, *J. Phys. Chem. Solids* 1983, **44**, 335.

The ratio of the measured hyperfine constants is not exactly equal to the ratio of the nuclear g_N factors. This is because the hyperfine constants are very large and the high-field approximation is no longer valid. Accurate values of the hyperfine constants in this case cannot be directly measured from the spectra and have to be obtained using simulations.

Spin polarization

The unpaired electron in the methyl radical is in a p-orbital

Apart from σ-radicals, there is a large class of radicals that have unpaired electrons in p- or π-orbitals. These radicals are called π-radicals. For instance, the methyl radical has a trigonal planar geometry and hence the unpaired electron is in a pure p-orbital. As p-orbitals have zero density at the nucleus, the hyperfine interaction in the methyl radical cannot be explained by direct Fermi contact interaction. This interaction is accounted for in terms of *spin polarization.* The electrons in the vicinity of the unpaired electron will preferentially have the same spin orientation as the unpaired electron. This is similar to Hund's rule which states that in the absence of other factors, electrons prefer to have parallel spin. A simplistic visualization of this effect is shown in Fig. 4.15. Two arrangements of the electron spins in the C–H bond with respect to the unpaired electron of the methyl radical can be envisaged (Fig. 4.15). The arrangement in Fig. 4.15(a) is preferred.

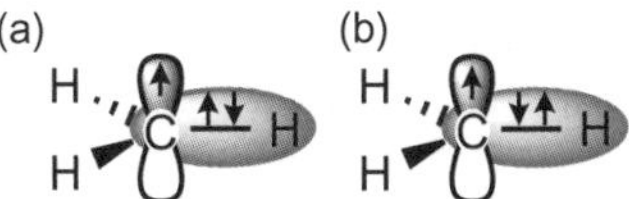

Fig. 4.15 Simplistic illustration of spin polarization in the methyl radical: the distribution of spin density in the C–H bond is polarized by the unpaired electron spin on the carbon atom.

This effect perturbs (i.e. *polarizes*, hence the term spin polarization) the distribution of spin density in the C–H orbital (sp^3-hybridized) which has some s-character and therefore some electron density at the H atom. The C–H bond thus mediates the Fermi contact interaction between the unpaired electron in the p-orbital and the H atom.

As spin polarization involves a small perturbation of the spin density, it is a much weaker effect than the direct Fermi interaction, and hence leads to much smaller hyperfine coupling constants. This can be used to determine the radical geometry. For instance, while a methyl radical has the unpaired electron in a p-orbital, a trifluoromethyl radical has tetrahedral geometry with the unpaired electron in an sp^3-hybridized orbital. This difference in geometry can be explained by electronegativity. Bonds with electronegative substituents tend to have higher p-character than bonds with electropositive ones (*Bent's rule*). As fluorine is much more electronegative than hydrogen, C–F bonds in the trifluoromethyl radical have higher p-character (sp^3) than C–H bonds in methyl radical (sp^2). Hence in a ^{13}C-labelled trifluoromethyl radical, there is a direct Fermi interaction with the ^{13}C nucleus. On the other hand, a ^{13}C-labelled methyl radical is a π-radical, and hyperfine interaction is only possible via spin polarization. As a result, the hyperfine constant $a(^{13}C)$ for the trifluoromethyl radical is nearly ten times larger than that for the methyl radical (Fig. 4.16).

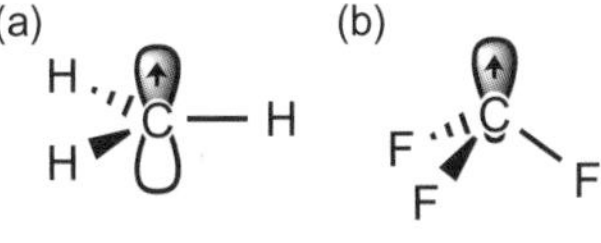

Fig. 4.16 Spin polarization and direct Fermi interaction in (a) the ^{13}C-labelled methyl and (b) trifluoromethyl radicals give rise to very different ^{13}C hyperfine constants.

Determining distribution of spin density: McConnell equation

In Chapter 2, eqn 2.20 shows that the value of the hyperfine coupling constant is proportional to the $|\psi(0)|^2$ term which is the probability of finding the unpaired electron at the nucleus. This relationship makes it possible to estimate the distribution of electron density in delocalized radicals. It is most commonly used for the carbon-centred radicals in the form of the empirical McConnell equation:

$$a = Q\rho \tag{4.1}$$

Here ρ is the unpaired electron density (so called *spin density*) at the carbon atom, *a* is the hyperfine constant for the α-proton, and *Q* is the proportionality coefficient. Empirically *Q* is close to 2.3 mT.

The value of Q depends slightly on the charge and structure of the radical. However, the approximate value of 2.3 mT is sufficiently accurate in most cases.

For instance, in any localized radical such as the methyl radical, the spin density at the carbon atom is unity, and hence hyperfine interaction with the adjacent protons (α-protons) is a = 2.3 mT × 1 = 2.3 mT, which is close to the experimental value (Fig. 4.17a). In fact, according to eqn 4.1, all carbon-centred localized radicals should have very similar hyperfine constants for α-protons.

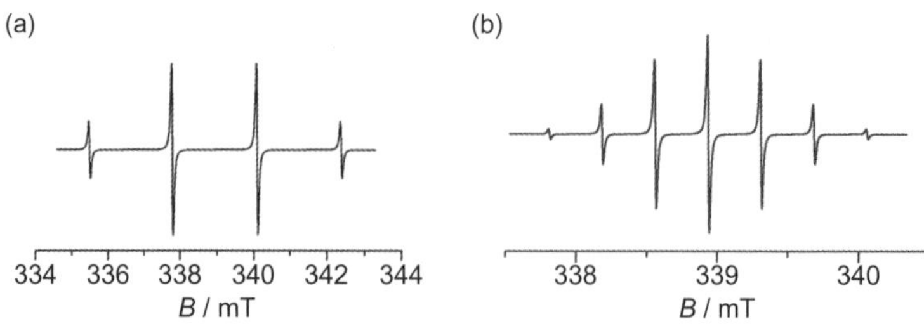

Fig. 4.17 EPR spectra of: (a) the localized methyl radical and (b) the delocalized benzene radical anion. The hyperfine constants for the α-protons in the two radicals are 2.3 and 0.375 mT, respectively.

Hyperfine constants and spin densities could be either positive or negative (Chapter 2, section 2.4). The sign of the hyperfine constant cannot be determined directly from EPR spectra, as only the absolute value of the hyperfine constant is measured. Therefore, one has to treat eqn 4.1 with caution. For instance, while the total spin density of a monoradical (e.g. the sum of spin densities on all atoms) is exactly 1, the sum of *absolute values* of spin densities may be greater than one (as some spin densities can be positive, and some negative). If the sign of the hyperfine constant is included in eqn 4.1, the value of Q is –2.3 mT.

The sign of the hyperfine constant is determined by the relative energies of the m_I levels. It is usually determined using theoretical calculations or special advanced EPR techniques.

The McConnell equation is very important for delocalized radicals. For instance, in the benzene radical anion, the unpaired electron is uniformly distributed among all six carbon atoms. The spin density at each carbon is therefore 1/6, and the hyperfine constant for coupling of the unpaired electron with each hydrogen is $a = 2.3/6 = 0.38$ mT, again, very close to the experimentally determined value (Fig. 4.17b).

Historically, the expansion of EPR in the 1950s followed the development of Hückel MO theory in the 1930s (which predicts the distribution of electron density in conjugated systems). As the experimentally determined hyperfine constants are linearly proportional to spin densities, EPR spectroscopy at the time was one of the most direct methods to test the results of the calculations, and it facilitated acceptance of the Hückel theory by the wider community. Table 4.1 shows a few examples of how the calculated spin densities compare to the hyperfine constants determined from EPR data.

Table 4.1 Experimental ^{1}H hyperfine constants a(H) and spin densities ρ calculated using the McConnell equation and Hückel theory

Radical anion	a(H)/mT	ρ (McConnell)	ρ (Hückel)
Benzene	0.375	0.163	0.167
Naphthalene	0.501 (α), 0.179 (β)	0.218 (α), 0.078 (β)	0.181 (α), 0.069 (β)
Butadiene	0.762 (α), 0.279 (β)	0.332 (α), 0.122 (β)	0.362 (α), 0.138 (β)

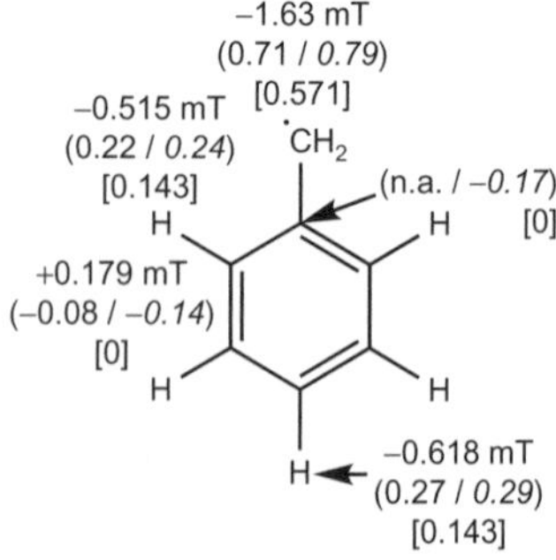

Fig. 4.18 Experimental hyperfine constants and spin densities calculated using eqn 4.1 (round brackets), modern DFT theory (in italics in round brackets), and Hückel theory (square brackets) for the benzyl radical.

Applications of the McConnell equation are not limited to radical anions. The assigned EPR spectrum of the benzyl radical (Fig. 4.12) makes it possible to measure the hyperfine constants, which for the α-, ortho-, meta-, and para-protons are 1.63, 0.515, 0.179, and 0.618 mT, respectively. One can easily see that the spin densities (calculated by dividing the hyperfine constants by the empirical value of $Q = 2.3$ mT) do not add up to one. This is because meta-hyperfines are positive, whereas the other hyperfines have negative sign. The results do not match Hückel theory well, as the latter predicts zero spin density for the meta-carbon atoms (Fig. 4.18). This discrepancy highlights the shortcomings of the Hückel theory. Modern quantum chemistry software packages (e.g. freeware ORCA, https://orcaforum.cec.mpg.de or commercial software Gaussian, http://www.gaussian.com) make it possible to accurately calculate the distribution of spin densities and hyperfine constants (Fig. 4.18).

Determining radical geometry: β-hyperfine constants

We have seen earlier that β-hyperfine constants in π-radicals can be larger than α-hyperfines (Fig. 4.8). This is explained by *hyperconjugation*, caused by overlap of the p-orbital bearing the unpaired electron with the sp^3-orbital of the C–H bond at the adjacent carbon atom (Fig. 4.19a). This overlap is most efficient when the dihedral angle θ between the p-orbital with the unpaired electron and the C–H bond is zero. This is easiest to see in a Newman projection (Fig. 4.19b).

In the first approximation, the value of the hyperfine constant for a β-H is proportional to the spin density ρ on the α-carbon and $\cos^2\theta$:

$$a = B\rho\cos^2\theta \qquad (4.2)$$

where B is the proportionality coefficient. Empirically, $B \approx 5.4$ mT. Together with the McConnell equation, this relationship makes it possible to determine the radical geometry and the distribution of electron density in different organic radicals.

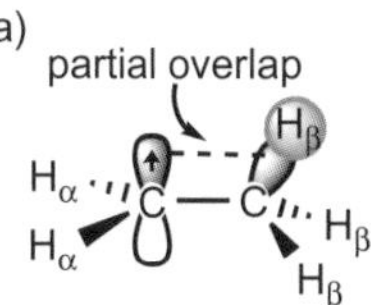

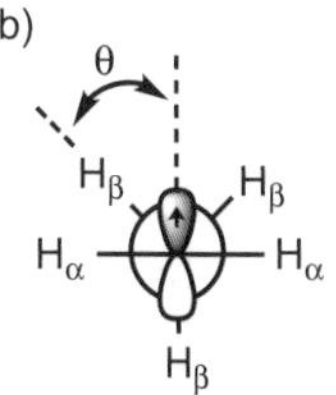

Fig. 4.19 Hyperconjugation in π-radicals: (a) flying wedge and (b) Newman projections.

In many acyclic radicals, the rotation across the C_α–C_β bond is very fast. This leads to the averaging of the dihedral angle θ over the full circle. The effective angle $\cos^2\theta$ in the case of free rotation is 0.5. Therefore the hyperfine splitting for the β-protons in localized radicals (ρ = 1) is $a = B \times 1 \times ½ = 2.7$ mT if there is unrestricted rotation across the C_α–C_β bond. For instance, the propyl radical (Fig. 4.8) has hyperfine values of $a = 2.2$ and 2.9 mT, in reasonable agreement with the values predicted via eqns 4.1 and 4.2 for the α- and β-protons (2.3 and 2.7 mT, respectively). All localized carbon-centred radicals with unrestricted rotation arcoss the C_α–C_β bond should have β-hyperfine values around 2.7 mT.

Average value of a function $f(x)$ over interval [0, a] is $\frac{1}{a}\int_0^a f(x)dx$. The effective $\cos^2\theta$ in the case of free rotation is thus $\frac{1}{2\pi}\int_0^{2\pi}\cos^2\theta d\theta = 0.5$.

Equation 4.2 is particularly powerful in predicting the geometry of cyclic radicals and other radicals with fixed geometry. In the absence of free rotation, the β-hyperfine constant can be quite large (up to the value of $B = 5.4$ mT) and much larger than α-hyperfines. For instance, consider the cyclohexyl radical (Fig. 4.20a). At low temperature, the equatorial and axial β-protons give different hyperfine splittings as the ring inversion is slow on the EPR timescale. The experimentally determined hyperfine values ($a(H_\alpha) = 2.13$, $a(H_\beta^{ax}) = 3.94$, and $a(H_\beta^{eq}) = 0.53$ mT, Fig. 4.20c) can be used with eqn 4.2 to give dihedral angles for the equatorial and axial protons of 84° and 23°, respectively, in good agreement with the values predicted for the perfect chair geometry ($\theta^{eq} = 90°$ and $\theta^{ax} = 30°$, Fig. 4.20b). At higher temperatures, the peaks for axial and equatorial protons coalesce and spectrum analysis makes it possible to calculate kinetic parameters for the ring inversion (see section 8.3).

'Slow on the EPR timescale' means that the difference between hyperfine constants for different conformations (expressed in frequency units) is much larger than the frequency of conformational change. The difference between axial and equatorial hyperfine constants is 3.41 mT or 95.6 MHz. See section 8.3 for more details.

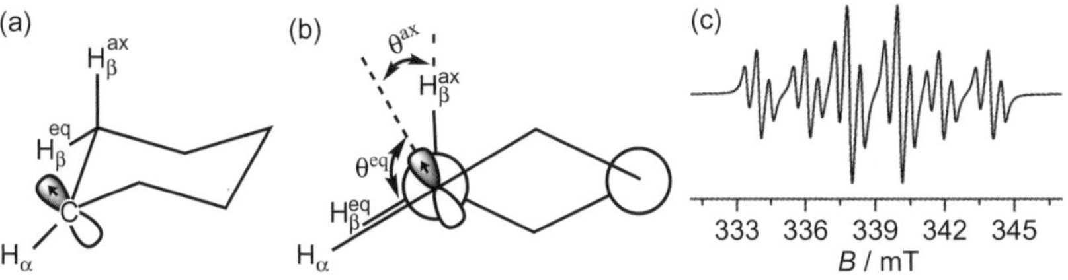

Fig. 4.20 The structure of the (a) cyclohexyl radical, (b) its Newman projection, and (c) the resulting EPR spectrum.

After Y. Hori, S. Shimada, and H. Kashiwabara, *J. Phys. Chem.* 1986, **90**, 3073.

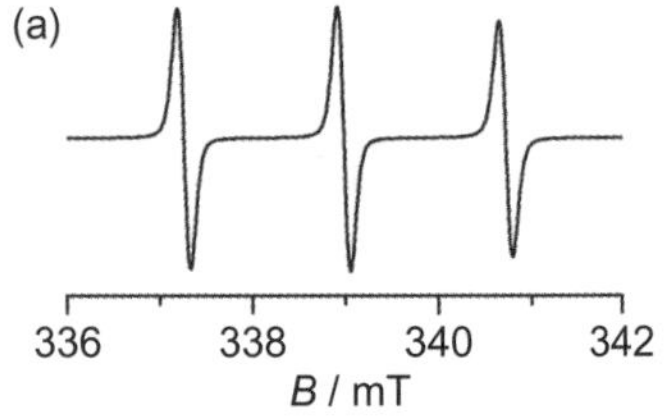

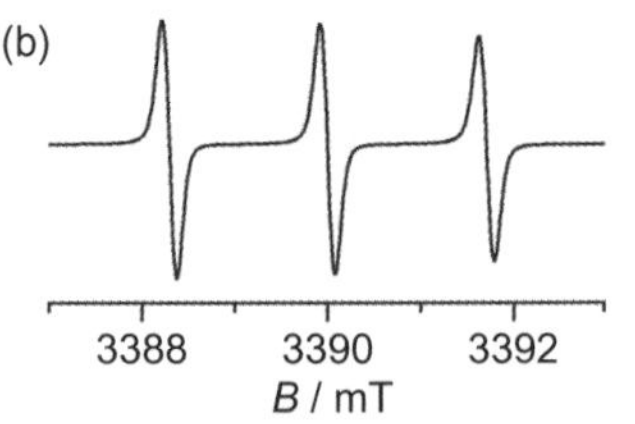

Fig. 4.21 EPR spectra of TEMPO at (a) X (9.5 GHz) and (b) W (95 GHz) bands.

After L. J. Berliner, Ed., *Spin Labeling: The Next Millennium*, Kluwer, 2002, p. 96.

EPR spectra of organic radicals at other frequencies

The EPR spectroscopy of organic radicals in solutions does not usually benefit from higher field/frequency (Fig. 4.21). The hyperfine interaction (eqn 2.12) is field-independent, and therefore higher field/frequency does not improve the resolution of the hyperfine patterns. The sensitivity improvement is partially negated by the need to use smaller sample volumes (see section 3.2). Therefore, the EPR studies of organic radicals in solution are best performed at X-band (9.5 GHz). The only tangible advantage of using higher frequency spectrometers for such samples is if a mixture of radicals with different g factors is studied. As the resonance field scales linearly with frequency (eqn 2.9), the spectra of two radicals with different g factors resolve better at higher frequency.

4.6 Generation and detection of organic free radicals

Free radicals are often considered as highly reactive species. However, this is not always the case. There are a large number of stable and persistent molecular radicals. The radical is defined as *stable* if it can be isolated as a pure compound; *persistent* radicals are loosely defined as those with a relatively long lifetime (longer than several seconds) under conditions that they were generated. Apart from the well-known examples of O_2 (which is a diradical, see Fig. 7.6), NO, and NO_2 (see Figs. 6.15 and 6.17), there are several classes of persistent and stable organic radicals, e.g. nitroxides, triarylmethyls, verdazyls, galvinoxyl, and others (Fig. 4.22). The stability of most of these

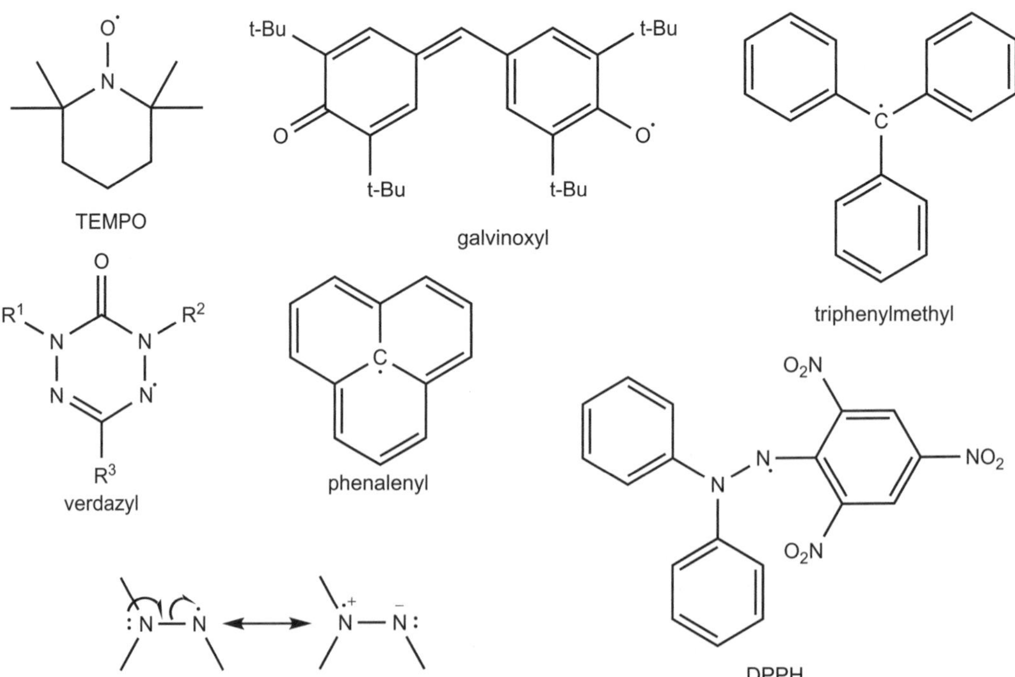

Fig. 4.22 Structures of some stable and persistent organic radicals, and radical stabilization of hydrazyl radicals.

radicals is explained by a combination of steric shielding (e.g. three phenyl rings in the triphenylmethyl radical prevent its dimerization), resonance stabilization (Fig. 4.22 shows one example for the hydrazyl radicals), and weak bonds in some dimers (e.g. nitroxides like TEMPO do not dimerize due to the weak O-O bond that would be formed in a dimer).

The range of reactivities of free radicals is truly staggering. Some species (such as the hydroxyl radical $^{\bullet}OH$) are extremely reactive and their lifetime in solutions is often determined by the rate of diffusion, usually measured in nanoseconds. Other radicals classed as reactive have much longer lifetimes. For instance, the superoxide radical, $O_2^{\bullet-}$, can have a lifetime of a few seconds in solution; this is at least nine orders of magnitude (!) longer than the hydroxyl radical.

The reactivity of free radicals determines the methods for their generation and detection. Although EPR spectroscopy is a sensitive technique (e.g. compared to NMR), it is not always possible to generate high enough steady state concentrations of the free radicals to enable direct detection. In this section, common radical generation/detection techniques are described.

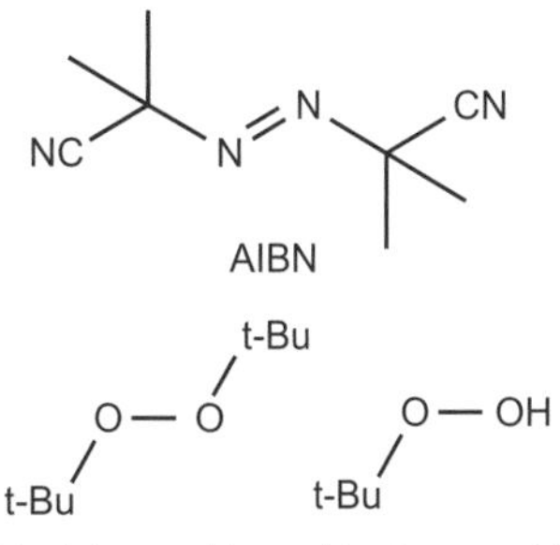

AIBN

t-butyl peroxide and hydroperoxide

Fig. 4.23 Examples of radical initiators.

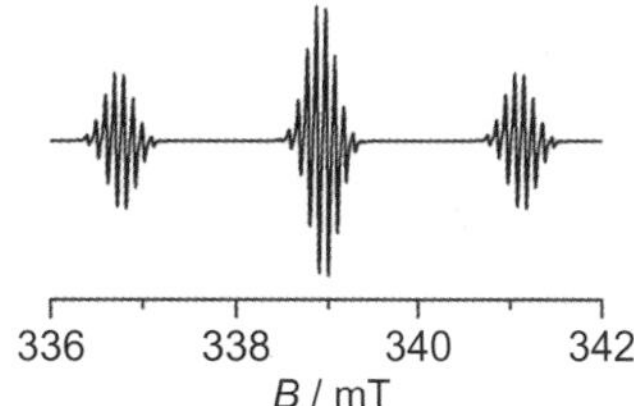

Fig. 4.24 EPR spectrum of the t-$BuCH_2^{\bullet}$ radical formed by photolysis of t-butylacetyl peroxide ([t-$BuCH_2CO]_2O_2$). Note a small hyperfine splitting from nine γ-protons.

After J. K. Kochi and P. J. Krusic, *J. Am. Chem. Soc.* 1969, **91**, 3940.

Radical generation

Radicals are produced by either homolytically breaking a weak bond, or in a single electron transfer process. The simplest way of breaking a bond is by heating, and compounds with weak bonds (e.g. organic peroxides, RO-OR, hydroperoxides, RO-OH, and azo compounds, $R_3C{-}N{=}N{-}CR_3$, Fig. 4.23) can be used to generate free radicals. These compounds are well known and commonly used as initiators of free radical polymerizations.

Peroxy and azo initiators also produce free radicals upon UV irradiation directly in the EPR cavity. The primary radicals generated from the initiators then undergo predictable chemical reactions, and this can be used to prepare the desired radicals. For instance, carboxylate radicals formed by irradiation of diacyl peroxides, release CO_2 and form alkyl radicals (Fig. 4.24).

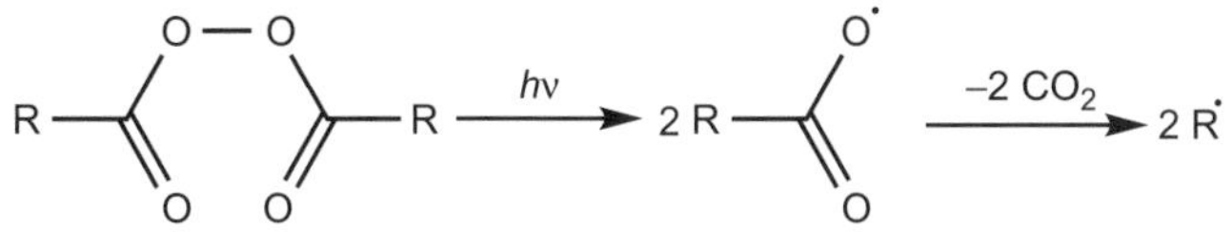

$$RC(O)O{-}OC(O)R \xrightarrow{h\nu} 2\,RC(O)O^{\bullet} \xrightarrow{-2\,CO_2} 2\,R^{\bullet}$$

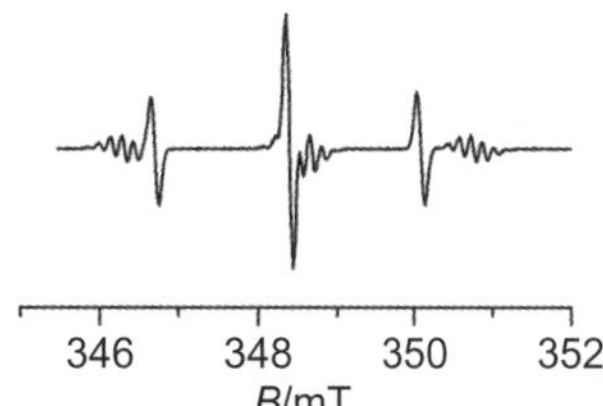

Fig. 4.25 EPR spectrum of the mixture of two radicals formed from MTBE. Radical A gives a triplet (large triplet in the spectrum). Radical B gives a triplet of heptets (a small hyperfine is due to six γ-protons).

Alkoxyl radicals generated by the homolytic cleavage of the O-O bond in organic peroxides, abstract hydrogen atoms from the weakest C-H bond which is useful for generation of stabilized carbon-centred radicals (Fig. 4.25).

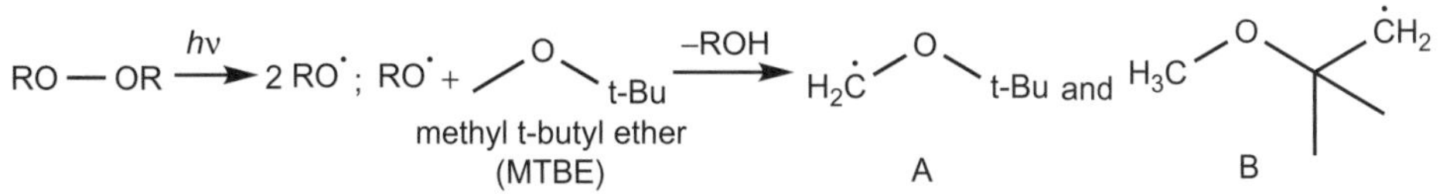

$$RO{-}OR \xrightarrow{h\nu} 2\,RO^{\bullet};\; RO^{\bullet} + CH_3{-}O{-}t\text{-}Bu \xrightarrow{-ROH} H_2\dot{C}{-}O{-}t\text{-}Bu \text{ and } H_3C{-}O{-}C(CH_3)_2\dot{C}H_2$$

methyl t-butyl ether (MTBE) — A — B

Other high energy radiation sources (e.g. ionizing radiation) or high energy electrons can also be used to produce free radicals. These methods were instrumental in the generation and characterization of a large number of hydrocarbon radicals in the early days of EPR spectroscopy.

Single electron transfer processes usually produce charged radicals (e.g. radical cations or radical anions) although they can be protonated or deprotonated to form neutral radicals. Common ways of achieving single electron transfer are electrochemical methods or reactions with redox-active transition metals. For instance, electrochemical reduction of electron-deficient aromatics yields corresponding radical anions.

$$\text{PhCN} \xrightarrow{+e^-} [\text{PhCN}]^{\bullet -}$$

Radical cations of aromatic amines could be produced by electrochemical oxidation. Electrochemistry is usually done directly in a special EPR cell equipped with working, counter and reference electrodes.

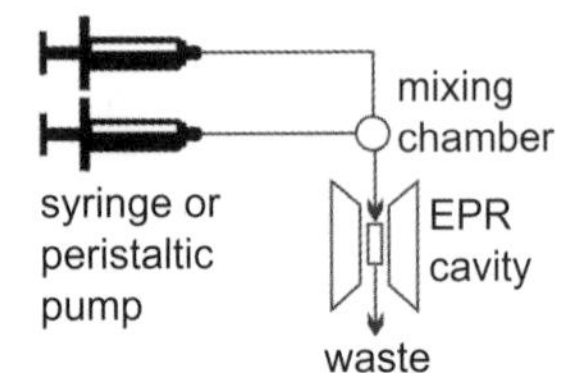

Fig. 4.26 A continuous flow EPR system.

All of the above methods (heating, irradiation, electrochemistry) enable continuous generation of radicals, often yielding steady state concentration which is sufficiently high for EPR detection. Chemical reagents (e.g. redox metals) are more difficult to add continuously. This is usually achieved with a continuous flow system (Fig. 4.26), where two or more reagents are pumped into a mixing chamber immediately above or below the EPR cavity. A common example of using this method for radical generation is the reaction of a redox active metal such as Fe(II) or Ti(III) with hydrogen peroxide in the presence of a suitable organic substrate. The Fenton reaction between the metal ion and H_2O_2 yields a hydroxyl radical that abstracts a hydrogen atom from the substrate producing a new radical that can be analysed by EPR. For instance, using ethanol as a substrate produces two ethanol radicals (Fig. 4.27):

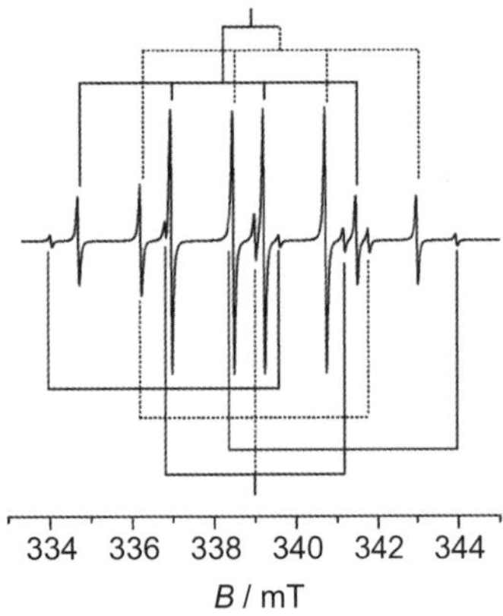

Fig. 4.27 The assigned EPR spectrum of ethanol radicals. The splitting diagrams for the major ($CH_3{}^{\bullet}CHOH$) and minor ($^{\bullet}CH_2CH_2OH$) components are shown above and below the spectrum, respectively.

$$H_2O_2 \xrightarrow{\text{Ti(III) catalysed Fenton reaction}} OH^{\bullet}$$

$$OH^{\bullet} + CH_3CH_2OH \xrightarrow{-H_2O} \dot{C}H_2CH_2OH + CH_3\dot{C}HOH$$

Many radical cations and anions are relatively long lived. Apart from resonance stabilization, this is explained by electrostatic repulsion between radical ions of the same charge, which disfavours their recombination. Thanks to relatively high stability, they can be prepared without the use of flow systems. Apart from electrochemical methods, radical ions are often prepared by reduction with alkali metals (e.g. radical anions of aromatic compounds) or oxidation in strong Brønsted or Lewis acids (radical cations).

$$C_6H_6 + K \longrightarrow [C_6H_6]^{\bullet -}\ K^+$$

Spin trapping

Although a large number of free radicals have been prepared by the methods described in the previous section and studied by EPR, in many cases their transient concentration is too low for direct detection (concentrations above 1 μM are generally required for EPR detection at X-band). In particular, free radicals are common intermediates in chemical and biological reactions, but their direct detection can be challenging. In some biological experiments, rapid freezing of samples prevents further reactions and intermediate radicals can be studied by EPR in the frozen solutions. Alternatively, the *spin trapping* technique is used.

Spin trapping is based on the addition of free radicals to double bonds. This is a very rapid and high yielding reaction for most radicals. Addition of a short-lived radical to a compound with a double bond (the *spin trap*) produces another radical (the *spin adduct*). The main principle behind spin trapping is that by using specially designed spin traps, the spin adducts are *persistent* (e.g. have relatively long lifetime), and so accumulate to sufficiently high concentrations to enable direct EPR detection.

Two classes of spin traps are commonly used, namely nitroso derivatives and nitrones. Examples of commercially available spin traps are shown in Fig. 4.28. Both classes form persistent nitroxides as spin adducts.

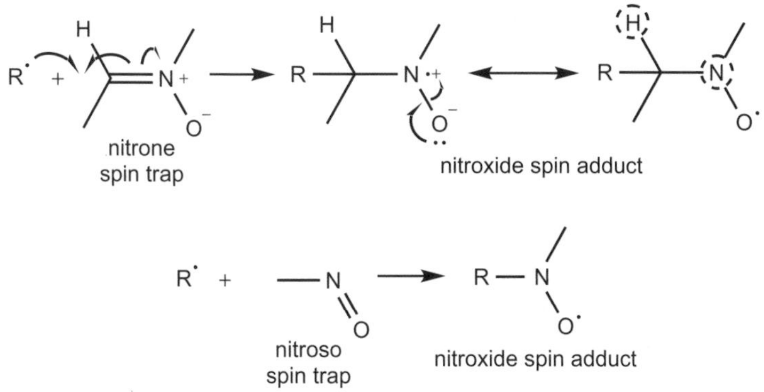

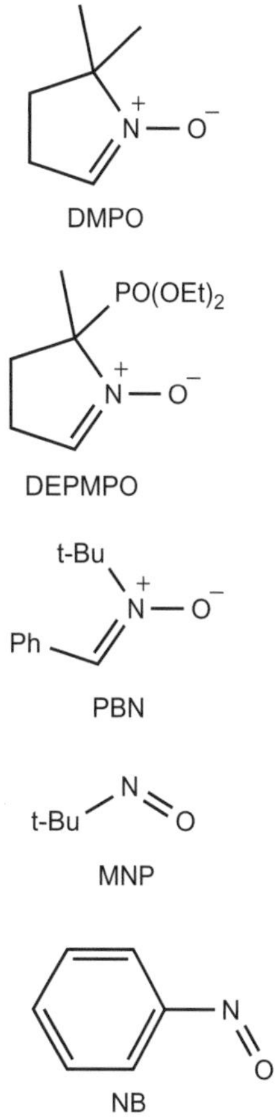

nitrone spin trap

nitroxide spin adduct

nitroso spin trap

nitroxide spin adduct

As there is spin density on the nitrogen atom in nitroxides (see resonance structure in the above scheme), the EPR spectra of spin adducts show hyperfine interactions with the nitrogen nucleus and any other nuclei with $I \neq 0$ attached to the nitrogen or one atom further away. In particular, the spin adducts of nitrones show hyperfine interactions with the nitrogen nucleus and β-protons (circled on the scheme above). These hyperfine constants are quite sensitive to the nature of the trapped radical $R^{\cdot}$. Although the relationship between the hyperfine constants and the structure of radical $R^{\cdot}$ is empirical, it often allows one to deduce structural information about the trapped radical. The hyperfine constants of spin adducts have been tabulated in several reviews; most conveniently these data can be accessed online using the spin trap database (see, for example, http://tools.niehs.nih.gov/stdb/).

Fig. 4.29 shows EPR spectra of the spin adducts of spin trap PBN and hydroxyl and methyl radicals. Although the basic splitting pattern is the same in both cases, the hyperfine constants are clearly different; matching their values with those in the spin trap database makes it possible to establish the structure of the trapped radical.

Fig. 4.29 shows that for *acyclic* spin traps such as PBN (Fig. 4.28), the sensitivity of the hyperfine constants to the structure of the trapped radicals is limited. The hyperfine constants of spin adducts of *cyclic* spin traps (such as DMPO, Fig. 4.28) are much more sensitive to the nature of the original radical. This is because β-hydrogen hyperfines strongly depend on the dihedral angle between the C–H bond and the orbital with the unpaired electron (section 4.5). In acyclic adducts the rotation around the C–N bond is almost unrestricted, and the dihedral angle averages. In the cyclic ones, the ring geometry is fixed, and the dihedral angle is determined by the most stable conformation, which depends strongly on the nature of the trapped radical (Fig. 4.30).

DMPO

DEPMPO

PBN

MNP

NB

Fig. 4.28 Examples of commercially available spin traps and their commonly used abbreviations.

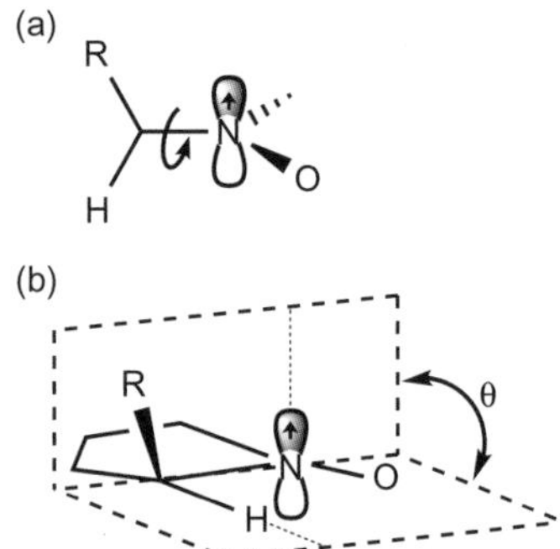

Fig. 4.30 (a) In acyclic spin adducts, free rotation around the C–N bond leads to the averaging of the dihedral angle between the C–H bond and the p-orbital hosting the unpaired electron and hence poor sensitivity of the hyperfine interactions to the nature of the trapped radical R. (b) In cyclic spin adducts, the ring geometry prevents free rotation around the C–N bond and the dihedral angle θ is very sensitive to the structure of radical R.

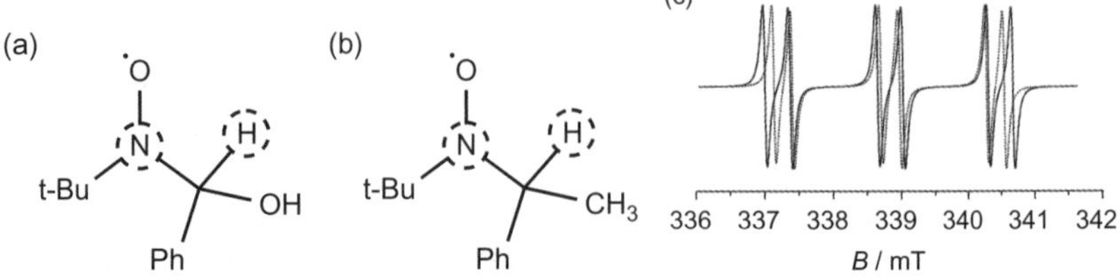

Fig. 4.29 The structure of (a) OH and (b) CH_3 adducts of the spin trap PBN and their corresponding EPR spectra. The spectrum of the OH adduct is shown with the dotted lines for clarity. The ^1H and ^{14}N nuclei participating in the hyperfine interaction are circled. The hyperfine constants are: $a_N = 1.57$ mT, $a_H = 0.27$ mT (OH adduct) and $a_N = 1.65$ mT, $a_H = 0.37$ mT (CH_3 adduct).

Although spin traps have allowed many transient radicals to be detected in the chemical and biological context, one should always exercise caution when interpreting the results of these experiments. Both nitroso- and nitrone-based spin traps are prone to artefacts, and the observation of spin adducts by EPR does not always imply the presence of free radicals. For instance, strong nucleophiles such as amines could add to nitrones to give hydroxylamines (by *Michael addition*) that spontaneously oxidize in air to form nitroxide spin adducts (Fig. 4.31a). The observation of EPR spectra of these adducts could lead to false conclusions about the presence of free radical intermediates in this system. Alternatively, strong oxidants (such as persulfate) oxidize nitrones to the corresponding radical cations which then rapidly undergo Michael addition with even weak nucleophiles to yield nitroxide spin adducts (Fig. 4.31b). Fortunately, careful control experiments in most cases make it possible to establish whether the spin adducts observed are due to the artefacts or genuine trapping of free radicals.

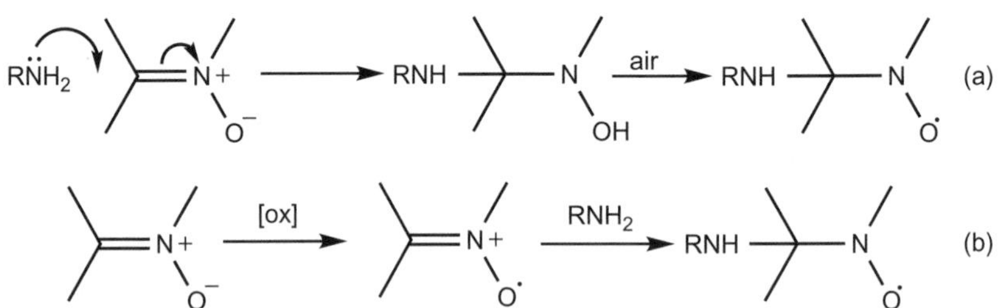

Fig. 4.31 Spin trapping artefacts leading to nitroxide spin adducts without trapping any radicals: (a) Forrester–Hepburn mechanism involving nucleophilic addition followed by oxidation (often by air), and (b) inverted spin trapping involving oxidation of the spin trap followed by nucleophilic addition.

4.7 Summary

- Hyperfine splitting of the EPR spectra of organic radicals in fluid solutions is dominated by the Fermi contact interaction. The resulting hyperfine patterns are invaluable for structure assignment.
- Hyperfine splitting is usually visible for nuclei attached to α- and β-atoms counting from the atom with the unpaired electron.

- The number of hyperfine lines follows the $2nI + 1$ rule, where n is the number of equivalent nuclei with nuclear spin I. For $I = ½$, the number and relative intensities of the lines can be determined using Pascal's triangle.
- The distance between the lines in a multiplet pattern equals the hyperfine constant.
- In π-radicals, the hyperfine interaction is explained by spin polarization. For α-protons in π-radicals, the hyperfine constant is linearly proportional to the spin density on the atom possessing the unpaired electron. For β-protons, the hyperfine constant is proportional to the spin density and $\cos^2\theta$, where θ is the dihedral angle between the p-orbital with the unpaired electron and the C–H bond.
- Short-lived free radicals are usually generated directly in the EPR cavity. If their steady state concentration is too low for detecting an EPR signal, the spin trapping technique can be used.

4.8 Exercises

4.1) Draw a splitting diagram for the cyclohexyl radical (Fig. 4.20) and assign the hyperfine pattern to the α- and β-protons.

4.2) Sketch an EPR spectrum of the deuterated propyl radical $CH_3CH_2CD_2^\bullet$ paying particular attention to the number of hyperfine lines, their relative intensity, and the separation between the lines. The EPR spectrum of the propyl radical is shown in Fig. 4.8.

4.3) Predict the number of lines and their relative intensity in the EPR spectrum of the cycloheptatrienyl radical ($C_7H_7^\bullet$). Estimate the hyperfine value for this radical and sketch the appearance of its EPR spectrum.

4.4) The EPR spectrum of the radical of glycolic acid (HOOC–CH$^\bullet$–OH) shows four lines with two hyperfine constants of 17 and 2.5 G. The unpaired electron is mostly localized on the carbon atom with some unpaired electron density found on the oxygen. Deduce which hydrogen atom has the larger hyperfine constant (17 G) and estimate the spin density on the carbon and oxygen atoms. Suggest the geometry of the glycolic acid radical consistent with the given hyperfine values.

4.5) Fig. 4.32 shows an EPR spectrum recorded when a solution of diphenylphosphine (PPh_2H) and PBN spin trap was treated with lead dioxide (PbO_2). Assign the EPR spectrum and deduce the structure of the intermediate radical formed.

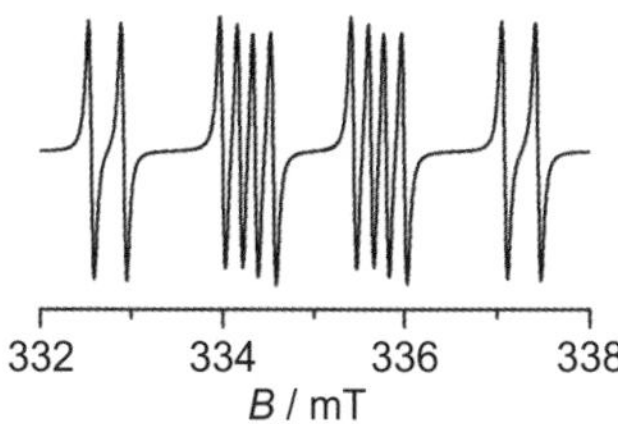

Fig. 4.32 EPR spectrum of the spin adduct (question 4.5).

5 Anisotropic EPR spectra in the solid state

5.1 Introduction

Rather than treating EPR theory from a classic perspective of charged particles interacting with a magnetic field, a quantum mechanics treatment of EPR is facilitated by use of the *spin Hamiltonian*. This contains the necessary terms to describe the spatial operators (such as spin and orbital angular momentum operators), including the field dependent and field independent operators, and will be used in this chapter. See bibliography for further reading.

In Chapter 4 the isotropic EPR spectra of spin doublets ($S = ½$) was presented. These spectra arise in fluid solution when the tumbling rate of the radical is much faster than the timescale of the EPR measurement (see section 8.3). In such conditions, only the *isotropic* spin Hamiltonian parameters (such as g_{iso} and a_{iso}) can be obtained. However, EPR spectra of paramagnetic centres can also be recorded in the solid phase, as single crystals, solid powders or frozen solutions. In the latter two cases, a random collection of orientations of the paramagnetic centres with respect to the applied magnetic field occurs resulting in *polycrystalline* or *powder* EPR spectra. Whilst these solid state spectra may be more difficult to analyse, they can also be more informative. The corresponding *anisotropic* **g** and **A** *tensors* (as opposed to the isotropic g_{iso} and a_{iso} values) extracted from the spectra provide additional information about the symmetry, bonding, structure and electronic configuration of the paramagnetic centre.

In this chapter, the origins of the anisotropies in **g** and **A** for a spin doublet ($S = ½$) are explored. We start by explaining how symmetry derived anisotropies in the solid state are manifested through **g** and how the interpretation of this tensor provides valuable information on the symmetry of the paramagnetic centre. The latter part of the chapter then focusses on the lineshapes for powder spectra and the origins of the hyperfine **A** tensor.

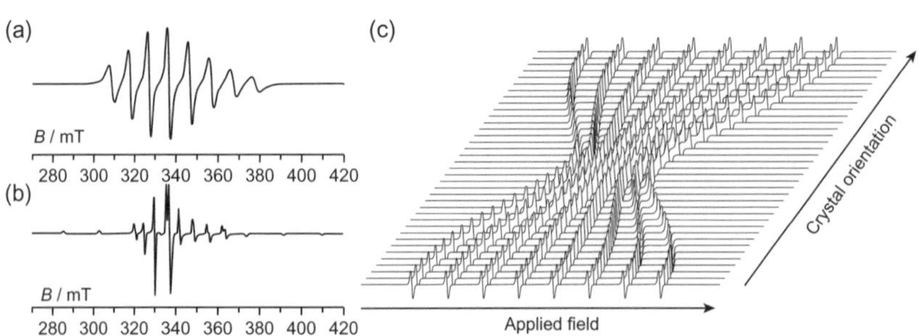

Fig. 5.1 CW EPR spectra of a VO^{2+} ion ($S = ½$, $I = 7/2$) recorded in different phases: (a) solution, (b) frozen-solution, and (c) dilute single crystal at different orientations. This figure illustrates the complexities of the solid state spectra (powder or single crystal) versus the solution EPR spectra.

To appreciate why it is necessary to understand the origins of the **g** and **A** tensors, consider the EPR spectra of a paramagnetic vanadyl (VO^{2+}) complex presented in Fig. 5.1. Since ^{51}V has a nuclear spin of $I = 7/2$, eight lines are expected, arising from the hyperfine interaction between the unpaired electron in the d^1 transition metal ion and the central vanadium nucleus. In the liquid phase, only the isotropic g_{iso} and a_{iso} spin Hamiltonian values can be extracted. However, in the solid state, the result of anisotropy in both **g** and **A** on the spectra are evident; the shift in the centre of the eight-line multiplet as the crystal is rotated arises from **g** *anisotropy*, whereas the variation in the line separation is caused by **A** *anisotropy* (Fig. 5.1). This chapter presents the theory explaining the origins of these anisotropies. Chapter 6 subsequently applies this theory to transition metal ions and inorganic radicals.

5.2 The anisotropic g tensor

In Chapter 2, the g value, is presented as a simple scalar quantity (i.e. the numerical factor linking spectrometer frequency to the magnitude of the applied magnetic field B in the resonant eqn 2.9) and independent of any orientation of the magnetic field vector $\boldsymbol{B}$ with respect to the paramagnetic centre. In Chapter 4, the g values are all close to free spin (g_e). However, in many chemical systems, the g values may deviate significantly from free spin. This is primarily due to the fact that the interaction of the electron spin with $\boldsymbol{B}$ is now orientationally dependent, caused by the localization of the electron within a particular orbital which introduces *orbital angular momentum* into the spin Hamiltonian equation. Spin–orbit coupling (SOC) is therefore the actual source of anisotropy in g. For this reason, the g factor must be replaced by a second rank **g** tensor to fully represent the orientational dependencies of the anisotropic interactions.

Just like a scalar or vector, a tensor is a mathematical object describing a physical property. The *rank* of the tensor depends on the number of directions needed to describe that property. If only one direction is needed, then the property can be defined by a 3 × 1 *column* vector. If two directions are required (e.g. as in EPR to represent the relative orientations of **g** vs **B**) then nine numbers are needed to define the property of **g**; these can be represented as a 3 × 3 symmetric matrix.

The anisotropy in the **g** tensor is usually classified in EPR as either *isotropic*, *axial*, or *rhombic*. The first class, *isotropic*, is rarely encountered in the solid state, and occurs only in systems possessing cubic, octahedral or tetrahedral symmetries and all other symmetries higher than cubic. As all three mutually orthogonal axes are equivalent in these cases, then all three g values will be equal ($g_{xx} = g_{yy} = g_{zz}$) and independent of the orientation of the paramagnet centre within the crystal with respect to $\boldsymbol{B}$. Systems possessing *axial* symmetry, for example with square pyramidal or trigonal pyramidal structures, have one unique axis. As a result, two g values will be equal whilst one will be different ($g_{xx} = g_{yy} \neq g_{zz}$). Finally, for systems with the lower *rhombic* symmetry (low point group symmetry), all three axes will be different, and this will produce three different g values ($g_{xx} \neq g_{yy} \neq g_{zz}$). This will be expanded further in section 6.3 (see Table 6.3).

The different g values arise due to variations in the orientation of $\boldsymbol{B}$ relative to the molecular frame and thus to the orientation of the paramagnet itself. To appreciate this, one must consider the different coordinate axes encountered in EPR analyses.

The anisotropic **g** tensor, or **g** parameter matrix, is often represented as a 3 × 3 symmetric matrix, dependent on the orientation of $\boldsymbol{B}$ with respect to g_{ij} in an arbitrary orientation. For an arbitrary orientation of g_{ij} relative to the laboratory or crystal axes, the matrix has the form:

$$\mathbf{g} = \begin{bmatrix} g'_{xx} & g'_{xy} & g'_{xz} \\ g'_{yx} & g'_{yy} & g'_{yz} \\ g'_{zx} & g'_{zy} & g'_{zz} \end{bmatrix}$$

In the principal axes (x, y, z) of the **g** tensor, the six off diagonal elements ($g_{yx}, g_{zx}, g_{zy}, g_{xy}, g_{xz}, g_{yz}$) are zero. This is referred to as a *diagonal tensor*:

$$\mathbf{g} = \begin{bmatrix} g_{xx} & 0 & 0 \\ 0 & g_{yy} & 0 \\ 0 & 0 & g_{zz} \end{bmatrix}$$

(see later note on the **gg** tensor).

g_{iso} is the average of the anisotropic g_{xx}, g_{yy} and g_{zz} values:

$$g_{iso} = \frac{g_{xx} + g_{yy} + g_{zz}}{3}$$

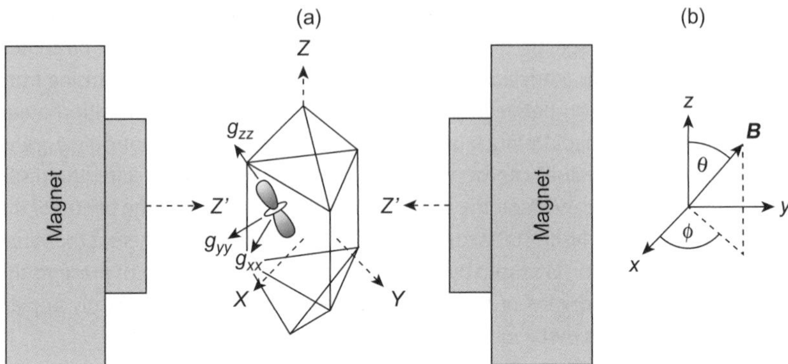

Fig. 5.2 (a) Schematic illustration of the three independent axes systems relevant to the EPR analysis, including the *magnetic field axes* (*X*′, *Y*′, *Z*′) the *principal* **g** *axes* (*x*, *y*, *z*) and the *crystal axes* (*X*, *Y*, *Z*). (b) In solid state systems we usually only consider the orientation of the external applied magnetic field ***B*** with respect to the principal orthogonal **g** tensor axes (*x*, *y*, *z*), using the right handed Cartesian coordinates with polar angles θ and ϕ.

By convention, the reference frame in EPR is usually taken to be the **g** tensor axes, rather than **A**.

These include the laboratory or applied *magnetic field axes* (labelled *X*′, *Y*′, *Z*′) the sample or *crystallographic axes* (labelled *X*, *Y*, *Z*) and the *principal* **g** *axes* (labelled *x*, *y*, *z*) of the paramagnet which we seek in the EPR analysis (Fig. 5.2). The **g** axes need not necessarily be orthogonal or coincident with the crystallographic axes (as illustrated in Fig. 5.2).

For axial symmetry ($g_{xx} = g_{yy} \neq g_{zz}$), the two equivalent *g* values are labelled '*g perpendicular*' ($g_{xx} = g_{yy} = g_{\perp}$) while the unique *g* value is labelled '*g parallel*' ($g_{zz} = g_{\parallel}$).

Since the anisotropies in the EPR spectrum when $I = 0$ arise principally from the energies of the electron spin interacting with the applied field, we need only to consider the orientations of **g** and ***B*** (Fig. 5.2) defined using Cartesian coordinates and two polar angles θ and ϕ. For the axial case, alignment of ***B*** along the unique *z* axis produces a resonance corresponding to one *g* value (i.e. correctly labelled g_{zz}), while alignment of ***B*** along the equivalent *x*,*y* axes will produce a second resonance or *g* value (i.e. g_{xx}, g_{yy}).

The origins of the g tensor

Direction cosines (symbol *l*) are used to specify the orientation of a vector (e.g. the vector components $B_{x,y,z}$) within a Cartesian coordinate frame where $l_x = \sin\theta\cos\phi$, $l_y = \sin\theta\sin\phi$, and $l_z = \cos\theta$. The normalization condition is $l_x^2 + l_y^2 + l_z^2 = 1$. Three *Euler angles* can also be used to specify an orientation.

As introduced above, the variation of ***B*** with respect to the principal **g** axes (expressed in the molecular frame of the paramagnetic centre) produces different experimental values of *g* that are observed in the EPR spectrum. Thus we must replace the simple resonant equation (eqn 2.9), with one which considers these angular terms, as given below:

$$h\nu = \mu_B g(\theta,\phi)B \tag{5.1}$$

The superscript T refers to the transposition of a row vector into a column vector. For example:

$$\mathbf{A} = \begin{bmatrix} a & b & c \end{bmatrix}$$

$$\mathbf{A}^{\mathrm{T}} = \begin{bmatrix} a \\ b \\ c \end{bmatrix}$$

where μ_B is the Bohr magneton, *B* is the strength of the applied magnetic field, and $h\nu$ is the energy difference between levels E_1' and E_2' (defined in Fig. 2.3). We must now consider the significance of the term $g(\theta, \phi)$ in eqn 5.1 and derive a more suitable form of the spin Hamiltonian that considers any arbitrary orientation of ***B*** relative to **g**. A more complete equation describing the interaction energy between the electron spin and ***B*** in an arbitrary coordinate system for the electron Zeeman Hamiltonian is given by:

$$\hat{H} = \mu_B \boldsymbol{B}^{\mathrm{T}} \cdot \mathbf{g} \cdot \hat{\boldsymbol{S}} \tag{5.2}$$

In this equation, $\hat{\mathbf{S}}$ is the Pauli spin operator vector with individual components $\hat{S}_x$, $\hat{S}_y$, $\hat{S}_z$. Since $\mathbf{B}$ can be aligned along any of the principal $\mathbf{g}$ tensor axes (as shown in Fig. 5.2 for an arbitrary orientation), then the individual components of $\mathbf{B}$ (labelled $B_{x,y,z}$) must be defined in terms of the polar angles (θ, ϕ), Fig. 5.3:

$$B_x = B\sin\theta\cos\phi \quad B_y = B\sin\theta\sin\phi \quad B_z = B\cos\theta \tag{5.3}$$

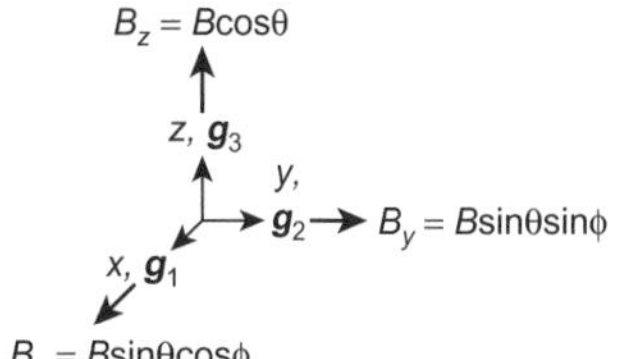

Fig. 5.3 $B_{x,y,z}$ refers to the components of the external field $\mathbf{B}$ aligned along the x, y, and z principal $\mathbf{g}$ axes.

Inserting the values of $\hat{S}_x$, $\hat{S}_y$, $\hat{S}_z$, and $B_{x,y,z}$ into eqn 5.2, and considering only the three diagonal elements of $\mathbf{g}$ (i.e. assuming the off-diagonal elements are zero), gives a more detailed matrix representation of the earlier spin Hamiltonian equation (eqn 5.2) given as:

$$\hat{H} = \mu_B \cdot \begin{bmatrix} B_x & B_y & B_z \end{bmatrix} \cdot \begin{bmatrix} g_{xx} & 0 & 0 \\ 0 & g_{yy} & 0 \\ 0 & 0 & g_{zz} \end{bmatrix} \cdot \begin{bmatrix} \hat{S}_x \\ \hat{S}_y \\ \hat{S}_z \end{bmatrix} \tag{5.4}$$

Inserting eqn 5.3 into eqn 5.4, and after manipulation of this last equation (i.e. mathematically this is the product of a *row vector* multiplied by a *square matrix* multipled by a *column vector*) gives the following Hamiltonian:

$$\hat{H} = \mu_B \left(B\sin\theta\cos\phi \cdot g_{xx} \cdot \hat{S}_x + B\sin\theta\sin\phi \cdot g_{yy} \cdot \hat{S}_z + B\cos\theta \cdot g_{zz} \cdot \hat{S}_z \right) \tag{5.5}$$

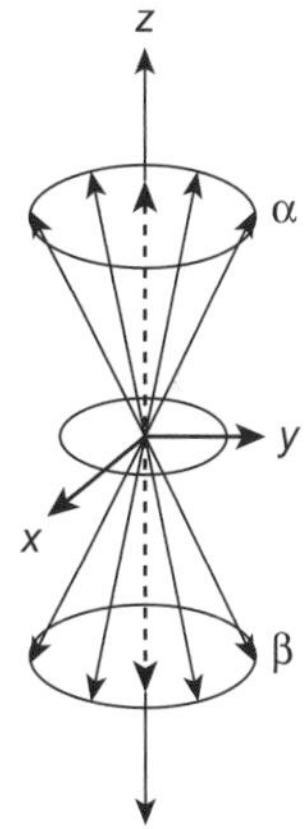

Fig. 5.4 Representation of the electron spin angular momentum.

Next, we must examine the significance of the $\hat{S}_x$, $\hat{S}_y$, $\hat{S}_z$ operators in more detail. As presented in eqn 2.1, the electron has intrinsic spin angular momentum. This property of spin is an observable quantity and the quantum mechanical operators for the spin components along the three orthogonal axes are given by the 2×2 *Pauli spin matrices* quantized in units of $\hbar$ as:

$$\hat{S}_x = \frac{1}{2}\begin{bmatrix} 0 & 1 \\ 1 & 0 \end{bmatrix} \quad \hat{S}_y = \frac{1}{2}\begin{bmatrix} 0 & i \\ -i & 0 \end{bmatrix} \quad \hat{S}_z = \frac{1}{2}\begin{bmatrix} 1 & 0 \\ 0 & -1 \end{bmatrix} \tag{5.6}$$

The key point to note from the above equations is that the matrix components of these values (0, ±1, ±*i*) are different along the three orthogonal *x*, *y*, and *z* directions (Fig. 5.4). The *z* component is shown in Fig. 5.4 with a dotted arrow, either pointing up (½ × 1) or pointing down (½ × −1). By comparison, the components in the *x*, *y* direction are not defined and therefore $\hat{S}_x$ and $\hat{S}_y$ have values of ½ × 0 in these two directions.

In order to find the solutions to eqn 5.5, we must consider the terms given in eqn 5.6 with eqn 5.5. The matrix of the Hamiltonian then has the form:

$$\hat{H} = \mu_B \frac{1}{2} B \begin{bmatrix} g_{zz}\cos\theta & g_{xx}\sin\theta\cos\phi + i g_{yy}\sin\theta\sin\phi \\ g_{xx}\sin\theta\cos\phi - i g_{yy}\sin\theta\sin\phi & -g_{zz}\cos\theta \end{bmatrix} \tag{5.7}$$

and the two resulting energy eigenvalues from this last equation become:

$$E_1 = -\frac{1}{2}\mu_B B \left(g_{xx}^2 \sin^2\theta\cos^2\phi + g_{yy}^2 \sin^2\theta\sin^2\phi + g_{zz}^2 \cos^2\theta \right)^{1/2} \tag{5.8a}$$

$$E_2 = \frac{1}{2}\mu_B B \left(g_{xx}^2 \sin^2\theta\cos^2\phi + g_{yy}^2 \sin^2\theta\sin^2\phi + g_{zz}^2 \cos^2\theta \right)^{1/2} \tag{5.8b}$$

For resonance absorption to occur, one must satisfy the relationship in the resonant equation (eqn 5.1) which states:

$$E_2 - E_1 = h\nu = g(\theta, \phi)\mu_B B \tag{5.9}$$

As a result the value of $g(\theta, \phi)$ in eqn 5.1 now becomes:

$$g^2(\theta, \phi) = g_{xx}^2 \sin^2\theta\cos^2\phi + g_{yy}^2 \sin^2\theta\sin^2\phi + g_{zz}^2 \cos^2\theta \tag{5.10}$$

Equation 5.10 gives the effective anisotropic g value for any orientation (θ, ϕ) with respect to the applied magnetic field $\boldsymbol{B}$.

5.3 The qualitative representation of $g(\theta, \phi)$

The angular terms in eqn 5.10 set limits on the magnitude of the g values, and thus the final **g** tensor, depending on how the paramagnetic centre is aligned or oriented. To understand the significance of this, we will next consider the profiles of the EPR spectra in the single crystal and powder cases.

The single crystal case

As mentioned above, the anisotropic **g** tensor in EPR can be classified as *isotropic*, *axial* or *rhombic*. The isotropic case ($g_{xx} = g_{yy} = g_{zz}$) only occurs for high symmetry centres. It is possible for some paramagnetic centres with lower symmetry elements to produce quasi isotropic spectra ($g_{xx} \approx g_{yy} \approx g_{zz}$) at low frequency (e.g. X-band). In such cases, higher field measurements are necessary to determine the accurate g values (see Chapter 9). As the orientation of $\boldsymbol{B}$ varies with respect to **g**, no change will be observed in the resonant field position B_{res}, since g remains an effective scalar or isotropic quantity.

For axially symmetric symmetry, the individual g values are not all equal ($g_{xx} = g_{yy} \neq g_{zz}$). The resonant field ($B_{res}$) at which microwave absorption occurs now depends on the angle between the applied field $\boldsymbol{B}$ and g_{zz} (or $g_{\parallel}$) aligned along the unique z axis. The effective g value for any orientation is then given by the positive square root of:

$$g^2(\theta) = g_{\perp}^2 \sin^2\theta + g_{\parallel}^2 \cos^2\theta \tag{5.11}$$

where θ is the angle between $\boldsymbol{B}$ and the z axis (Fig. 5.2b). Eqn 5.11 is a special case of eqn 5.10 where $\phi = 0°$ (i.e. the g values are independent of orientations in the x,y plane defined by ϕ). The two extreme or limiting g values occur at $g_{\parallel}$ (when $\boldsymbol{B}$ is parallel to the z axis, $B_{\parallel}$) and $g_{\perp}$ (when $\boldsymbol{B}$ is perpendicular to the z axis, $B_{\perp}$). This is illustrated in Fig. 5.5a showing the variation in $g(\theta, \phi)$ and resonant field (B_{res}) as θ changes from 0° to 180° (in this case, as the single crystal bearing the paramagnetic centre is rotated). It is important to realize that changes in the crystal orientation with respect to $\boldsymbol{B}$, will produce variations in $g(\theta, \phi)$ according to eqn 5.11 (i.e. upper spectra in Fig. 5.5a for orientations of 0°, 90°, and 180°). However, since the **g** axes (x, y, z) are the usual reference frame in EPR, it is more convenient to visualize the paramagnetic single crystal as fixed, with the applied field $\boldsymbol{B}$ varying.

According to eqn 5.9, the resonant field of the angular dependent axial case (Fig. 5.5a) is given as:

$$B_{res} = \frac{h\nu}{g(\theta)\mu_B}$$

In Fig. 5.5a $g_{\parallel} = 2.050$ when $\theta = 0°$ and $g_{\perp} = 2.002$ when $\theta = 90°$. The shape of this angular curve is represented by an effective g value given by the positive square root of eqn 5.10, where $[g^2(\theta)]^{1/2} = g(\theta)$.

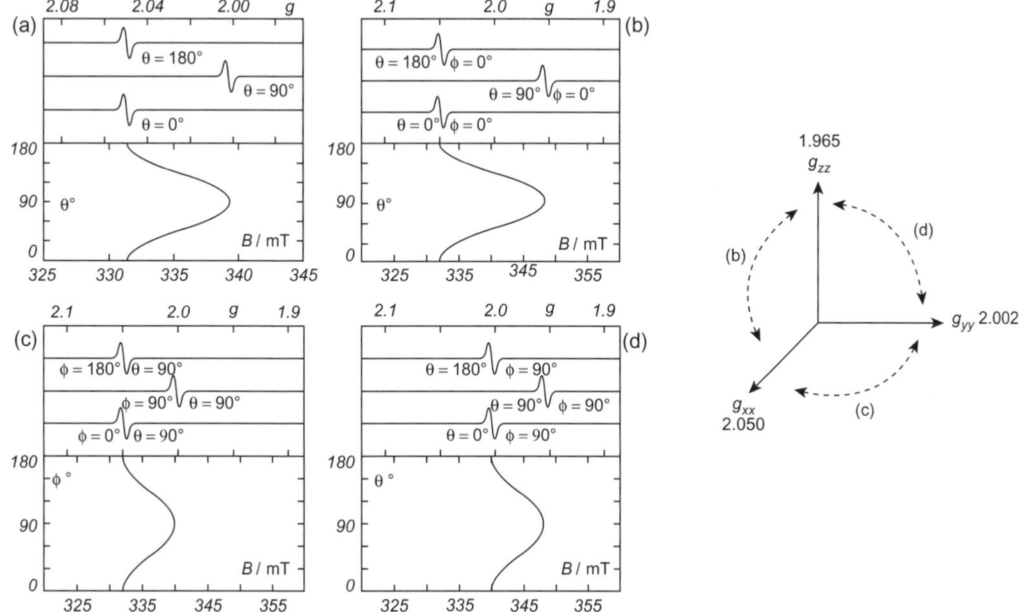

Fig. 5.5 Plots of variation in g values as a function of sample orientation, and associated angular variation in resonant field (B_{res}) as a function of the angles θ, ϕ. The *axial case* is illustrated in (a) for g values of $g_{\parallel} = 2.050$ and $g_{\perp} = 2.002$. The *rhombic case* is illustrated in (b–d) for g values of $g_{xx} = 2.050$, $g_{yy} = 2.002$, and $g_{zz} = 1.965$. In (b), ϕ is fixed at 0° (and only θ varies), while in (c) θ is fixed at 90° (while ϕ now varies in the *x,y* plane), and in (d) ϕ is fixed at 90° (with θ varying). The plane of rotation with respect to the **g** tensor is illustrated in the Cartesian coordinate frame on the right.

For systems possessing rhombic symmetry, the g values are all inequivalent ($g_{xx} \neq g_{yy} \neq g_{zz}$). The resonant fields and the effective g values for any arbitrary orientation are given once again as the positive square root of eqn 5.10. The resulting angular dependences of the g values are shown in Fig. 5.5(b–d). In Fig. 5.5(b), ϕ remains fixed at 0 while θ changes from 0 through 90° to 0 (i.e. from 0° to 90° to 180°) such that the g value progressively changes from g_{zz} to g_{xx} and back again to g_{zz}. The angular curve looks similar in Fig. 5.5(c, d) but one must note that the magnitude of the g values are different. In Fig. 5.5(c), ϕ = 90° while θ changes from 0° through 90° to 180°, such that the g value now changes from g_{zz} to g_{yy} and back again to g_{zz}. In Fig. 5.5(d), **B** is aligned along the *x,y* axes (θ = 90°); rotation of **B** in this plane from ϕ = 0° to 180°, causes a change from g_{yy} to g_{xx} and back again to g_{yy}.

Experimental determination of the g tensor

The process of systematically recording the single crystal EPR spectra for different orientations of the sample in the applied field is the manner used to extract the experimental g^2 values. This experiment and subsequent analysis is not easy. The reader interested in the underlying detailed methodology is directed to relevant excellent texts explaining the process and mathematics involved. However, a brief explanation can be given here to provide a qualitative understanding of the approach.

The single crystal is mounted with one axis (say the Y axis) vertical to the laboratory magnetic field, and using a goniometer, rotation about this axis provides a series of spectra within the XZ plane (illustrated in Fig. 5.6). Any convenient axis of the

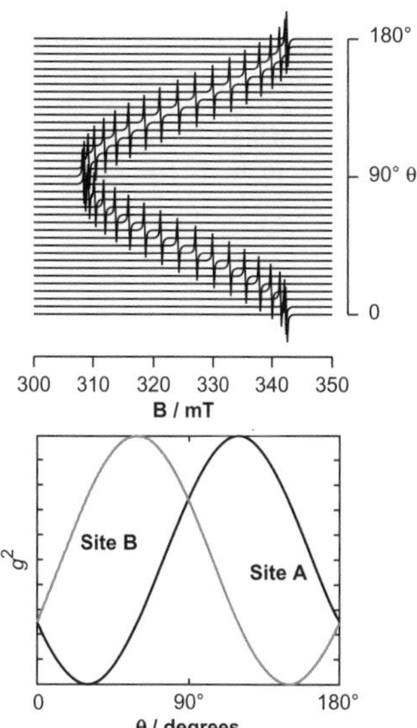

Fig. 5.6 Single crystal EPR spectrum for an $S = ½$ spin system in axial symmetry, showing the resulting angular dependency curve for g^2 as a function of θ following rotation in a single plane.

The **g** tensor is obtained by analysis of the experimental EPR spectrum from the observed $g^2(\theta, \phi)$ values using a single crystal.

crystal can be chosen for this measurement. A similar series of measurements are then repeated by rotating the crystal about the X or Z axes when they are aligned vertical to the magnetic field. Since the principal **g** axes of the paramagnetic centre are not necessarily coincident with the crystal axes (as illustrated in Fig. 5.2), one effectively obtains values of the elements of **g** in terms of the crystal axes and the off-diagonal elements will not be zero (the elements of **g** in this case are termed **gg**). If we measure the apparent g values when the magnetic field $\mathbf{B}$ has the components given in eqn 5.5, or in direction cosines Bl_x, Bl_y, Bl_z, then we can write:

Since the **gg** matrix is always symmetric (i.e. $g_{ij} = g_{ji}$), even if **g** is not necessarily symmetric across the diagonal, only the six components in the matrix are required (hence the lower three matrix elements are not shown in eqn 5.12).

$$g^2 = \begin{bmatrix} l_x & l_y & l_z \end{bmatrix} \cdot \begin{bmatrix} (\mathbf{gg})_{xx} & (\mathbf{gg})_{xy} & (\mathbf{gg})_{xz} \\ & (\mathbf{gg})_{yy} & (\mathbf{gg})_{yz} \\ & & (\mathbf{gg})_{zz} \end{bmatrix} \cdot \begin{bmatrix} l_x \\ l_y \\ l_z \end{bmatrix} \tag{5.12}$$

We may rotate the field in the XZ plane where the direction cosines are ($l_x = \sin\theta$, $l_y = 0$, $l_z = \cos\theta$). After inserting these field components into eqn 5.12, it follows that:

$$g^2 = (\mathbf{gg})_{xx}\sin^2\theta + 2(\mathbf{gg})_{xz}\sin\theta\cos\theta + (\mathbf{gg})_{zz}\cos^2\theta \tag{5.13}$$

noting that all terms containing y do not appear in this equation. The experimental curve of g^2 versus θ (Fig. 5.6) for this XZ plane is then fitted to eqn 5.13 to extract the three tensor elements $(\mathbf{gg})_{xx}$, $(\mathbf{gg})_{xz}$, $(\mathbf{gg})_{zz}$. The corresponding direction cosines for the YZ and XY planes are (0 sinθ cosθ) and (cosϕ sinϕ 0) respectively, so the measurements are repeated in these two planes and the resulting curves are fitted to the equivalent expressions as given in eqn 5.13 for YZ and XY. In this way, the tensor elements $(\mathbf{gg})_{yy}$, $(\mathbf{gg})_{yz}$, $(\mathbf{gg})_{zz}$ are extracted from the YZ plane and the elements $(\mathbf{gg})_{xx}$, $(\mathbf{gg})_{xy}$, $(\mathbf{gg})_{yy}$ from the XY plane. With information on the six independent $(\mathbf{gg})_{ii}$ components in eqn 5.12, one must finally find a matrix transformation which diagonalizes this tensor, such that the transformation will define the relative orientation of the crystal axes with respect to the principal **g** axes of the paramagnetic centre.

We can rotate a tensor matrix from the laboratory axes into the **g** axes (x,y,z). Since the **g** tensor is almost always symmetric ($g_{xy} = g_{yx}$, etc.), then diagonalization is achieved using the transformation $\mathbf{g}_{(\text{diag})} = \mathbf{R} \cdot \mathbf{g}' \cdot \mathbf{R}^{-1}$. The transformation matrix **R** defines the orientation of the new principal axes with respect to the old axes.

The powder case

Magnetically dilute single crystals can be difficult to prepare. Therefore EPR spectra are more frequently recorded as powders or dilute frozen solutions producing *polycrystalline* EPR spectra. The single crystal spectrum is unquestionably more informative; but nevertheless, a considerable amount of information can still be obtained from the powder pattern.

These polycrystalline spectra usually have broad linewidths, because the observed resonances produce an envelope of all possible peaks (or g values) due to the weighted distribution of all resonance fields. In a polycrystalline environment, all orientations have the same probability and therefore the total intensity of the EPR spectrum is given by the sum of the contributions from each single molecular orientation in a unit sphere. Let us examine the shape of the powder pattern for a simple $S = ½$ spin system in axial and rhombic symmetries but with no hyperfine interaction ($I = 0$).

The effective g values can be expressed in terms of three principal values depending on the orientation of the paramagnetic centre in the applied field. The powder spectra can give these principal values directly, but not the *principal directions* of the **g** tensor with respect to the molecular axes. These can only be determined experimentally using a single crystal (i.e. when the crystallographic and molecular axes alignment in the field are known).

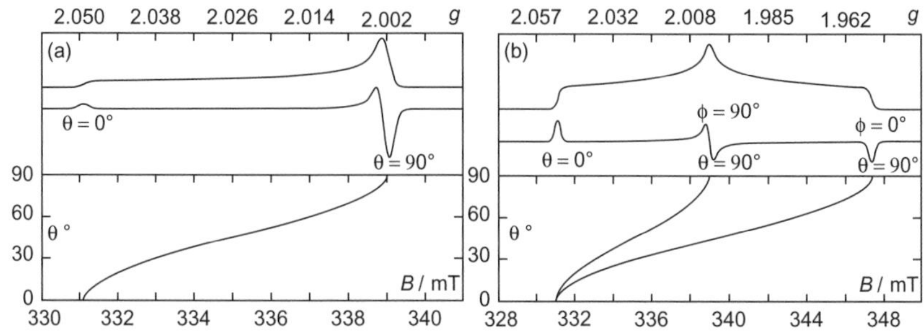

Fig. 5.7 Profile of the polycrystalline EPR spectra, in absorption mode and first derivative mode for an $S = ½$ spin system ($I = 0$) and associated angular variation in resonant field (B_{res}) as a function of the spherical polar angles θ, ϕ. (a) *Axial case* with $g_{\|} = 2.050$ and $g_{\perp} = 2.002$ and (b) *Rhombic case* with $g_{xx} = 2.050$, $g_{yy} = 2.002$, and $g_{zz} = 1.965$. In both cases, ν = 9.5 GHz.

The angular curve in Fig. 5.7(a) as a function of g can be conveniently calculated by:

$$\theta = \cos^{-1}\left[\frac{g^2(\theta) - g_{\perp}^2}{g_{\|}^2 - g_{\perp}^2}\right]^{½}.$$

For an axial $S = ½$ spin system, two principal g values exist ($g_{\perp}$ and $g_{\|}$). In this case, the variation in the g values depend solely on the vector orientation of ***B*** with respect to the unique z axis (defined by the angle θ). The values of $g_{\perp}$ and $g_{\|}$ will set the range of resonant fields (B_{res}) over which absorption occurs. When $g_{\|} > g_{\perp}$ then no absorption occurs at fields lower than $B_{\|}$ while no absorption occurs at fields higher than $B_{\perp}$:

$$B_{\|} = \frac{1}{g_{\|}}\frac{h\nu}{\mu_B} \qquad B_{\perp} = \frac{1}{g_{\perp}}\frac{h\nu}{\mu_B} \tag{5.14}$$

When $\theta = 0$ ($B_{\|}$) resonance absorption will occur only for those paramagnetic centres in which their z axis is aligned directly along the ***B*** direction (see Fig. 5.7a). At this orientation, very few spins contribute to the EPR line intensity. As the field moves progressively from $B_{\|}$ to $B_{\perp}$, more spins come into resonance and correspondingly the intensity of the absorption line increases. Eventually when θ = 90°, a maximum in absorption and therefore peak intensity is observed (see next section).

Although CW EPR spectra are recorded as first derivatives, the values of $g_{\perp}$ and $g_{\|}$ can still be extracted from the experimental powder pattern (Fig. 5.8). The angular dependency curve showing the variation in B_{res} with respect to θ is also shown in Fig. 5.7(a) (lower plot). Whilst the turning points in the spectrum can be readily identified for the two extreme limits, defined by θ = 0 and 90°, it is important to appreciate that a continuous absorption occurs across all field positions between $B_{\|}$ and $B_{\perp}$.

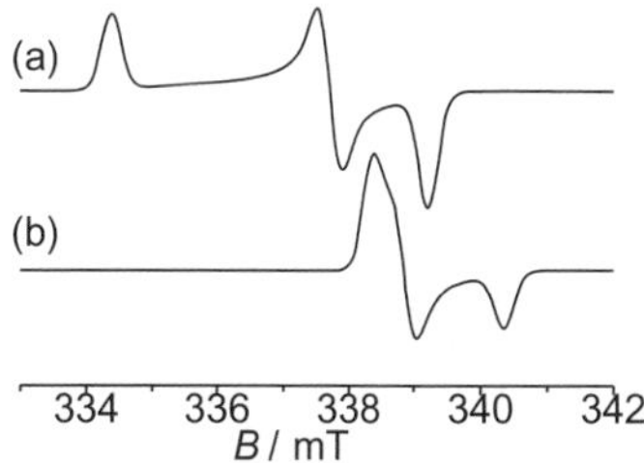

Fig. 5.8 Polycrystalline EPR spectra (a) the S_2O^- radical on MgO ($g_{xx} = 2.001$, $g_{yy} = 2.030$, $g_{zz} = 2.010$), and (b) a phenoxyl radical ($g_{xx} = 1.994$, $g_{yy} = 2.003$, $g_{zz} = 2.006$). Note the *quasi*-axial appearance of (b) due to the smaller difference in g_{ii} values compared to (a), where the rhombic profile is more evident at this measurement frequency. For (b) at this frequency, $g_{zz} = 2.006 \approx g_{yy} = 2.003$ ($g_{\perp}$) $> g_{xx} = 1.994$ ($g_{\|}$), unlike the examples in Fig. 5.9 where $g_{\|} > g_{\perp}$.

S_2O^- parameters after M. J. Lin and J. H. Lunsford, *J. Phys. Chem.* 1976, **80**, 635.

For a rhombic $S = ½$ spin system, the variation in the g values depends on the two polar angles of θ and ϕ (eqn 5.10). An example powder pattern for such a system is shown in Fig. 5.7(b). Three special cases or *singularities* occur for B_{res} corresponding to i) θ = 0, ii) θ = 90° = ϕ, and iii) θ = 90°, ϕ = 0. When θ = 0, the spins in resonance are those for which ***B*** lies along the z axis and an absorption edge occurs producing the derivative peak corresponding to g_{zz}. As the field moves away from the z axis, into the zy plane (such that ϕ = 90° and only the angle of θ varies) B_{res} will also vary and a maximum in intensity of the absorption occurs when θ = 90° = ϕ (g_{yy} in Fig. 5.7b). A similar situation occurs when the field moves from the z axis but now in the zx plane, such that all intermediate values of θ contribute to the intensity of the absorption line

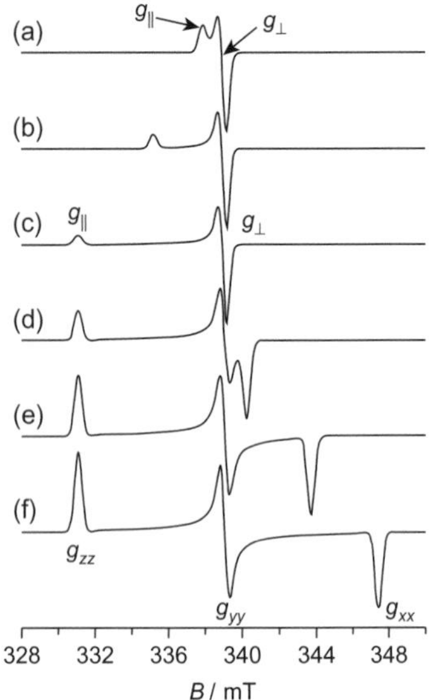

Fig. 5.9 Powder CW EPR profiles for axial and rhombic $S = ½$ spin systems, showing the variations in the shape of the spectra. To a first approximation, the turning points may be associated with the **g** tensor components. The discrepancy is caused by line broadening. These spectra represent the powder summed average of the individual orientations in the single crystal case.

Powder EPR spectra, with the correct associated peak intensities, can be very accurately reproduced using freely available EPR simulation packages such as *EasySpin* (see Chapter 4) and *EPRsim32* (see Bibliography).

Eqn 5.15 is only valid for axial systems; otherwise $d\phi$ must be taken into consideration for rhombic systems.

When $\theta = 0$, then $h\nu = g_{\|}\mu_B B_{\|}$. Thus:

$$\left(\frac{h\nu}{\mu_B}\right)^2 = \left(g_{\|}B_{\|}\right)^2$$

reducing the overall B dependency in eqn 5.16 to simply B^{-1}.

(since $\phi = 0$ in this plane). The limiting point for this trend is reached when $\theta = 90°$, $\phi = 0$ (g_{xx} in Fig. 5.7b).

The angular dependency curves illustrating the variation in B_{res} are shown in the lower plot in Fig. 5.7(b). It is important to note, that for these rhombic cases, while only very few orientations contribute to the spectrum when $\boldsymbol{B}$ is parallel to the g_{zz} axis and $\boldsymbol{B}$ is parallel to the g_{xx} axis (sometimes referred to as *single crystal-like* orientations), several intermediate orientations reveal the same resonance for the position where $\boldsymbol{B}$ is parallel to the g_{yy} axis resulting in a maximum absorption at this field.

The simulated powder EPR spectra for axial and rhombic $S = ½$ spin systems are shown in Fig. 5.9. For axial systems (Fig. 5.9a–c), the $g_{\|}$ component can be recognized as the peak which does not cross the base-line (unlike the perpendicular component which does), with $g_{\|} > g_{\perp}$. In Fig. 5.9(a), the **g** anisotropy is small ($g_{\|} \approx g_{\perp}$). As the anisotropy increases (due to an increase in axial distortion), the $g_{\|}$ values progressively shift to lower fields (Fig. 5.9b, c). For rhombic systems (Fig. 5.9d–e), three peaks are present. In Fig. 5.9(d), the system is quasi-axial ($g_{zz} \neq g_{yy} \approx g_{xx}$). As the extent of distortion in the *x,y* direction increases, the **g** anisotropy correspondingly increases (Fig. 5.9e–f) to eventually give $g_{xx} \neq g_{yy} \neq g_{zz}$.

Peak intensities in powder spectra

As shown in Fig. 5.7(a), for a system with axial symmetry, B_{res} depends on the g values ($g_{\|}$, $g_{\perp}$) and θ to produce the overall powder spectrum. To understand the relative intensities of the peaks in the powder spectrum, one must sum over all values of θ to obtain the correct lineshape and intensities.

In a statistically randomly oriented ensemble, all orientations of the spin with respect to $\boldsymbol{B}$ will be equally probable. The magnetic field vector $\boldsymbol{B}$ is defined using the Cartesian coordinate frame (Fig. 5.2b and Fig. 5.10). Since all orientations of the spin system are equally probable, some centres will be in resonance and contribute to the spectral intensity at a field between the two extremes $B_{\|}$ and $B_{\perp}$ for an *axial* system. It is then necessary to sample a sufficient number of orientations of $\boldsymbol{B}$ in the *x,y,z* directions in order to produce the correct powder lineshape. It can be shown, for axial systems only in absorbance mode, that the probability (P) of a spin system experiencing a resonant field between B and $B + dB$ is dependent on $\sin\theta$:

$$P(B) = C\frac{1}{2}\sin\theta\left(\frac{dB}{d\theta}\right)^{-1} \tag{5.15}$$

where C is a normalization constant. It is important to understand the significance of this relationship between $P(B)$ and $\sin\theta$. As seen in Fig. 5.10, a large number of centres will be present with symmetry axes lying in the *x,y* plane, compared to the smaller number aligned along the *z*-direction. In this sense, the probability of observing a resonance at the $g_{\|}$ orientation is very low. However, as CW EPR spectra are recorded in first derivative mode, the fastest changes are observed at the edges of the spectrum (Fig. 5.11) hence a peak appears at the turning angles of $B_{\|}$ and $B_{\perp}$. By taking the derivative of $dB/d\theta$ in eqn 5.15, one obtains:

$$P(B,\theta) = \left(\frac{h\nu}{\mu_B}\right)^2 \frac{1}{B^3\left|\left(g_{\perp}^2 - g_{\|}^2\right)\cos\theta\right|} \tag{5.16}$$

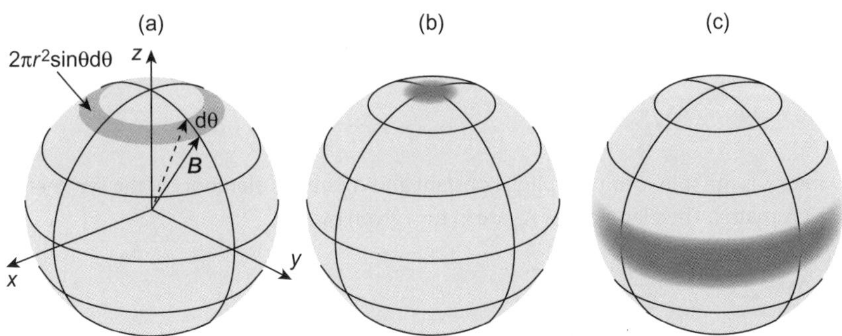

Fig. 5.10 Unit spheres representing the density of populations for (b) ***B*** parallel to *z* and (c) ***B*** perpendicular to *z*, and (a) for an arbitrary orientation of ***B***.

For an axial system B_{res} is given by:

$$B_{res} = \frac{h\nu}{g(\theta)\mu_B}$$

Expanding and simplifying gives:

$$B_{res} = \left[g_\perp^2 \sin^2\theta + g_\parallel^2 \cos^2\theta\right]^{-1/2} \frac{h\nu}{\mu_B}$$

$$B_{res} = \left[g_\perp^2 - \left(g_\perp^2 - g_\parallel^2\right)\cos^2\theta\right]^{-1/2} \frac{h\nu}{\mu_B}$$

Therefore some spin centres will always be in resonance at B_{res} positions between $B_\parallel$ and $B_\perp$.

This means that the probability of observing an EPR line is zero ($P(B) = 0$) outside the extremes of $g_\parallel$ and $g_\perp$ (i.e. outside the field range $B_\parallel$ and $B_\perp$) and between the two extremes this probability is proportional to B^{-1}. Since cosθ appears in the denominator of eqn 5.16, then when $\theta = 0$, a finite value of $P(B)$ will be obtained. As the angle θ systematically increases towards 90°, when the field is aligned along the *x,y* axes, $P(B)$ will also increase monotonically to infinity at $\theta = 90°$ (Fig. 5.11). The profile of the curve shown in Fig. 5.11 corresponds to the hypothetical case with negligible linewidth. In real systems, the linewidths are usually broadened, resulting in the absorption profile shown in Fig. 5.7. This is the reason why the $g_\perp$ peak has considerably more intensity compared to the $g_\parallel$ peak in powder spectra.

Finally, it is important to appreciate that linewidths are very different in liquid versus powder spectra (see Chapter 8). In the liquid phase, the paramagnetic centres will experience a *time-averaged* response, leading to narrow lines. By comparison, in powders or frozen solution the centres experience a *space-averaged* response. In other words, each centre has its own resonance position, so the entire spectrum will be broadened since these combined resonances produce an envelope resulting in a weighted distribution of all possible resonance fields.

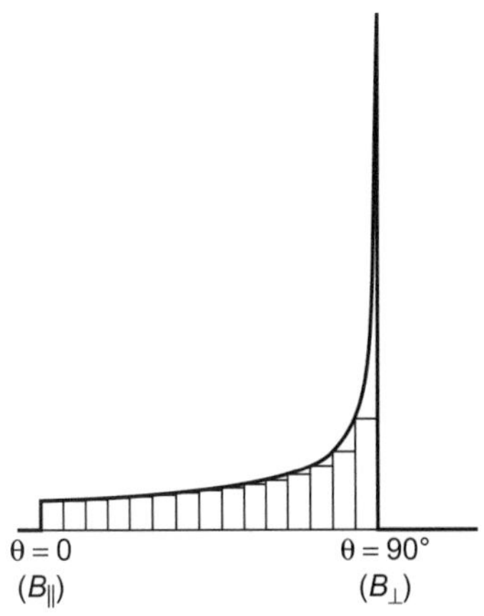

Fig. 5.11 Theoretical absorption curve for an axial system. The total intensity for an axial or rhombic case is given by the summed contributions for each molecular orientation within the unit sphere: $\int P(B)dB = \int_0^{\pi}\int_0^{2\pi} \sin\theta d\theta d\phi$. These powder patterns are usually computed numerically (using a simulation program) by systematically varying the angles θ and ϕ between 0 and 180° and 0 to 360° and weighting the spectral contributions with sinθ.

The influence of spin-orbit coupling on the g tensor

As introduced in Chapter 2, the electron possesses an intrinsic spin angular momentum which is responsible for the associated *g* factor ($g_e = 2.0023$). However, there are actually two contributions to the electron angular momentum that influence the *g* values; namely spin (***S***) and orbital (***L***) angular momentum which are coupled by spin-orbit coupling (symbol λ). Although ***L*** is usually suppressed or quenched in real systems with non-degenerate energy levels, λ restores some of the orbital momentum and this produces a further deviation of *g* from g_e. Up to this point we have only considered the effects of the electron spin (***S***) interacting with ***B***, and now we will examine how orbital angular momentum through spin-orbit coupling contributes to **g** anisotropy.

Spin-orbit coupling can admix ground states with excited states, causing a small amount of orbital angular momentum to appear in the ground state. This will result in a local magnetic field ($\boldsymbol{B}_{local}$) which adds to the external field (***B***) to produce shifts in the *g* values (Δg). The extent of this mixing depends on the energy difference between the ground and excited states, and which orbital (p, d, or f) contributes to

The total magnetic moment of the electron is the vector sum of the *spin* and *orbital* angular momenta: $\boldsymbol{\mu} = \mu_B(\boldsymbol{L} + g_e\boldsymbol{S})$.

In general, the ground and excited states in systems bearing heavy elements lie closer to each other compared to systems with lighter elements. As a result, the effect of spin-orbit coupling on EPR spectra of organic radicals (which consist of light elements only) is often negligible, but in transition metal ions it is often very significant.

Λ is defined as the spin-orbit effect perturbation matrix. The symmetry of the ion creates an anisotropic influence of the spin-orbit coupling on g.

$|0\rangle$ and $|n\rangle$ are the spin-orbit coupled electronic energies of the ground and excited states of energies E_0 and E_n.

Equations 5.17 and 5.18 are derived from *perturbation theory*.

the ground state. The g_{ij} components of the **g** tensor can then be expressed more formally as:

$$g_{ij} = g_e\delta_{ij} + 2\lambda\Lambda_{ij} \tag{5.17}$$

where λ is the spin-orbit coupling constant and Λ_{ij} are the elements of the symmetric 3×3 $\mathbf{\Lambda}$ matrix. The elements of Λ_{ij} are in turn given by:

$$\Lambda_{ij} = \sum_{n\neq 0} \frac{\langle 0|L_i|n\rangle\langle n|L_j|0\rangle}{E_0 - E_n} \tag{5.18}$$

Here L_{ij} are the orbital angular momentum operators for the x, y, or z directions. The Λ_{ij} elements account for the interaction of the ground state of energy E_0 with an excited state of energy E_n.

To understand the significance of this equation on the **g** tensor, consider the simple case of an unpaired electron in a p_z-orbital with the degeneracy shown in Fig. 5.12. The lowest lying p_z-orbital represents the doubly degenerate ground state (owing to the electron spin $S = \pm½$). This degeneracy is lifted in the presence of the applied magnetic field, which will interact with both the spin ($\hat{S}$) and orbital ($\hat{L}$) moments of the electron. Strictly speaking, the Hamiltonian then formally becomes:

$$\hat{H} = \mu_B \boldsymbol{B}^T\left(\hat{\boldsymbol{L}} + g_e\hat{\boldsymbol{S}}\right) + \lambda\hat{\boldsymbol{L}}^T\hat{\boldsymbol{S}} \tag{5.19}$$

although eqn 5.2 can be used instead of eqn 5.19 to extract the components of the **g** tensor using eqn 5.17. As a consequence, when the applied field $\boldsymbol{B}$ is aligned along the *x,y,z* direction (Fig. 5.12), a different set of g_{ij} values will be produced. When the field is oriented along the x-direction, the eigenvalues become:

$$E_x = \pm\frac{1}{2}\mu_B\left(g_e - \frac{\lambda}{\Delta_2}\right)B \tag{5.20}$$

One can gain a qualitative view of how the singly occupied orbital is transformed by a rotation about the direction axis of the applied field. In this case, when $\boldsymbol{B}$ is aligned along the x axis, spin-orbit coupling regenerates some orbital angular momentum between p_y and p_z. This mixing depends on the magnitude of λ and the energy separation Δ_2 between p_y and p_z (written as $E_0 - E_n$ above). The separation between the levels E_x becomes:

$$\Delta E_x = \mu_B\left(g_e - \frac{\lambda}{\Delta_2}\right)B \tag{5.21}$$

and as a result the resonance g_{xx} value is given by:

$$g_{xx} = g_e - \frac{\lambda}{\Delta_2} \tag{5.22}$$

Similarily when $\boldsymbol{B}$ is aligned along the y axis, orbital angular momentum is restored between p_x and p_z *via* λ with an energy separation of Δ_1. However, when $\boldsymbol{B}$ is aligned along the z axis, rotation of p_z does not transform it into one of the other p-orbitals (see section 6.2) so that g_{zz} does not shift from g_e:

$$g_{yy} = g_e - \frac{\lambda}{\Delta_1} \quad g_{zz} = g_e \tag{5.23}$$

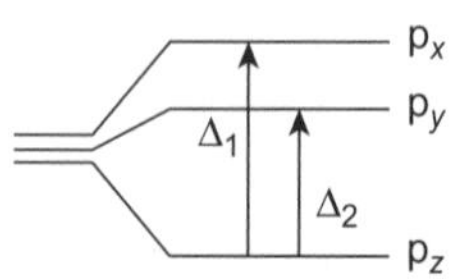

Fig. 5.12 Orbital splitting diagram illustrating the degeneracy of the p-orbitals in a site with rhombic symmetry.

The important point to note here is that λ and the energy separations $\Delta_{1,2}$ affect the g values, leading to anisotropy in **g**. Analysis of the g values can therefore provide important insights into the molecular orbitals hosting the electron. As a rule of thumb spin–orbit coupling to empty molecular orbitals produces negative Δg_{ij} shifts (i.e. $g < g_e$) whereas coupling to filled molecular orbitals produces positive Δg_{ij} shifts (i.e. $g > g_e$). This idea is developed further in Chapter 6.

5.4 The anisotropic hyperfine A interaction

As discussed in Chapters 2 and 4, the hyperfine interaction occurs when the unpaired electron interacts with the magnetic moment of a nearby nucleus of spin $I \geq \frac{1}{2}$. The two contributions to the hyperfine interaction are the isotropic Fermi contact term and the anisotropic dipolar term. If the magnitude of the hyperfine interaction is sufficiently large, then additional *hyperfine splittings* can be observed in the EPR spectrum. In the solid state, both isotropic and anisotropic components of the hyperfine will contribute to the EPR spectrum. The dipolar component of the hyperfine (labelled **T**) is particularly useful, providing information on the electron spin–nuclear spin distances and p-, d-, or f-orbital spin densities. Owing to the orientation of these orbitals with respect to the applied field, this dipolar term is anisotropic.

A_0 is the *isotropic hyperfine coupling constant* (units of Joules), measuring the magnetic interaction energy between the electron and the nucleus. This should not be confused with the anisotropic hyperfine values (A_{xx}, A_{yy}, A_{zz}) which are given in magnetic field or frequency units.

a_0 (the theoretical isotropic hyperfine value for unit spin density) or a_{iso} (the experimental isotropic hyperfine) is the *isotropic hyperfine splitting constant* expressed in frequency (A_0/h) or magnetic field ($A_0/(g\mu_B)$) units.

a_{iso} is the average of the anisotropic A_{xx}, A_{yy}, and A_{zz} values: $a_{iso} = \frac{A_{xx} + A_{yy} + A_{zz}}{3}$.

The anisotropic A tensor

In Chapter 2, the dipolar component of the hyperfine interaction was introduced briefly (section 2.4). The energy of this dipole–dipole interaction (E_{dip}) can be approximated to:

$$E_{dip} = -\frac{\mu_0}{4\pi}\left(\frac{3\cos^2\theta - 1}{r^3}\right)\mu_{elec}\mu_{nuc} \tag{5.24}$$

where μ_{elec} and μ_{nuc} are the components of the electron and nuclear dipole moments along ***B***, r is the distance between the two dipoles and θ is the angle between ***B*** and the vector joining the dipoles. The magnitude of E_{dip} clearly depends on r,θ. As shown in Fig. 5.13, the nucleus generates a local field $\boldsymbol{B}_{local}$ at the electron. Depending on the angle of θ, $\boldsymbol{B}_{local}$ will either add to or oppose ***B*** at the electron. In other words, the magnitude of the E_{dip} will be anisotropic.

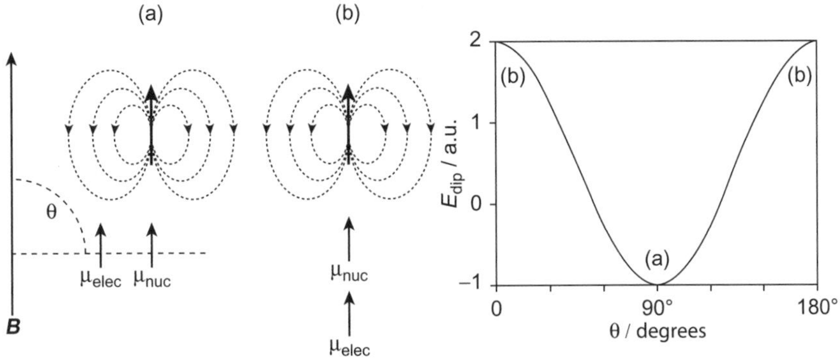

Fig. 5.13 Illustration of the magnetic field generated by the nucleus which creates a local field $\boldsymbol{B}_{local}$ at the electron. Depending on the angle of θ, $\boldsymbol{B}_{local}$ can either oppose or add to ***B*** with variations shown in (a) and (b). As a result E_{dip} will vary as a function of the angle θ from 0 to 90° to 180°. μ_{elec} and μ_{nuc} represent the location of the electron and nucleus, respectively.

In Fig. 5.13 the electron is shown to be localized as a fixed point. In reality this is not the case, and the electron will be delocalized so that E_{dip} must then be averaged over the entire electron probability distribution function. For a spherical distribution of electron density, all values of θ will be probable and the value of $\langle\cos^2\theta\rangle$ becomes 1/3. In this case, the $\boldsymbol{B}_{local}$ term vanishes and E_{dip} becomes zero. This situation arises when the electron is in a spherically symmetrical s-orbital, or when the paramagnetic centre is in the liquid phase undergoing rapid tumbling. For a non-spherical distribution of electron density, E_{dip} will be non-zero and the size of this energy term will depend on the orientation (i.e. it is anisotropic).

Taking the anisotropic hyperfine interaction into consideration, then the spin Hamiltonian for a single unpaired electron ($S = ½$) in an applied field and coupled to one or more nuclei of spin $I \geq ½$ becomes:

$$\hat{H} = \mu_B \boldsymbol{B}^T \cdot \mathbf{g} \cdot \hat{\boldsymbol{S}} + g_N \mu_N \boldsymbol{B}^T \cdot \hat{\boldsymbol{I}} + \hat{\boldsymbol{S}}^T \cdot \mathbf{A} \cdot \hat{\boldsymbol{I}} \tag{5.25}$$

The first and second terms corresponds to the electron Zeeman and nuclear Zeeman energies, while the third term describes the hyperfine interaction. **A** refers to the *hyperfine tensor* of the interacting nucleus and is composed of both an isotropic (a_{iso}) and pure anisotropic or dipolar part (**T**):

A is the hyperfine parameter matrix, whereas **T** is the purely *anisotropic hyperfine interaction matrix*, which is also represented as a 3 × 3 matrix (eqn 5.27).

$$\mathbf{A} = \begin{bmatrix} A_{xx} & A_{xy} & A_{xz} \\ A_{yx} & A_{yy} & A_{yz} \\ A_{zx} & A_{zy} & A_{zz} \end{bmatrix} = a_{iso} + \mathbf{T} \tag{5.26}$$

The pure dipolar component of the interaction energy, including the anisotropic contributions, is expressed by **T**:

$$\mathbf{T} = \begin{bmatrix} T_{xx} & T_{xy} & T_{xz} \\ T_{yx} & T_{yy} & T_{yz} \\ T_{zx} & T_{zy} & T_{zz} \end{bmatrix} \tag{5.27}$$

s-orbital occupation (C_s^2) or density (ρ^s) is given by a_{iso}/a_o when the electron is localized in an s-orbital.

In simple terms, p- or d-orbital occupation ($C_{p,d}^2$) or density ($\rho^{p,d}$) of the electron is given by T/b_o, where b_o is the theoretical anisotropic hyperfine value for *unit* spin density in a p- or d-orbital.

In this equation, **T** is a traceless, axially symmetric tensor with principal axes values of $T_{\parallel}$ and $T_{\perp}$ for a system where g is isotropic. Because the tensor is traceless, it results that $Tr(\mathbf{T}) = 0$ and therefore $T_{\parallel} = 2T_{\perp}$. In many real situations, both a_{iso} and **T** will contribute to the experimental hyperfine tensor (**A**), as for example in a radical with the unpaired electron localized in an sp-hybridized orbital. The spin density associated with the fractional s-orbital occupancy will then contribute to a_{iso}, while the dipolar **T** part will arise due to the fractional p-orbital occupancy. The principal components of the hyperfine $\mathbf{A}_{exp}$ can then be expressed for a simple axial case as:

$$\mathbf{A}_{exp} = \begin{bmatrix} A_{xx} & 0 & 0 \\ 0 & A_{yy} & 0 \\ 0 & 0 & A_{zz} \end{bmatrix} = \begin{bmatrix} a_{iso} - T & 0 & 0 \\ 0 & a_{iso} - T & 0 \\ 0 & 0 & a_{iso} + 2T \end{bmatrix} \tag{5.28}$$

According to eqn 5.28, the magnitude of the hyperfine coupling in the solid state will then range from $a_{iso} - T$ to $a_{iso} + 2T$, and the anisotropic coupling will change periodically as the angle or orientation of the paramagnetic species varies.

It is important to note, that for an electron localized in p- or d-orbital (for example a SOMO of predominant p- or d-character) then an axial hyperfine tensor should be observed. However, since the **T** tensor will depend on the radial and angular parts of the orbital hosting the unpaired electron, the elements of T_{ii} will formally be given by:

$$T_{ii} = \frac{\mu_0}{4\pi}\frac{g_e\mu_B g_N\mu_N}{r^3}\left\langle 3\cos^2\theta - 1\right\rangle \quad (5.29)$$

The term $g_e\mu_B g_N\mu_N r^{-3}$ is often abbreviated to P, where P_p represents unit population of the p-orbital and P_d = unit population of the d-orbital.

where θ is the angle between r and the applied field **B**. When the coefficients of the integrated orbital radial terms (see Bibliography for more details) are considered for p- or d-orbitals, then T_{ii} takes the general form:

$$\frac{2}{5}P_p\begin{pmatrix} 2 & & \\ & -1 & \\ & & -1 \end{pmatrix} \quad \text{and} \quad \pm\frac{2}{7}P_d\begin{pmatrix} 2 & & \\ & -1 & \\ & & -1 \end{pmatrix} \quad (5.30)$$

For example, if the SOMO is predominantly p-based: $|\mathrm{SOMO}\rangle = c_x|p_x\rangle + c_y|p_y\rangle + c_z|p_z\rangle$ then the dipolar matrix elements become: $T_{xx} = \frac{4}{5}c_i^2P_p$, $T_{yy} = -\frac{2}{5}c_i^2P_p$, $T_{zz} = -\frac{2}{5}c_i^2P_p$, where c_i are orbital coefficients.

In these situations, a non-diagonal **T** tensor may be observed, which must be diagonalized to find the principal values.

Hyperfine powder lineshapes

The lineshape for hyperfine splittings in polycrystalline media can be very difficult to deconvolute and interpret. Clearly, E_{dip} between the electron and the interacting nucleus must be considered for all possible orientations of the electron–nucleus vector

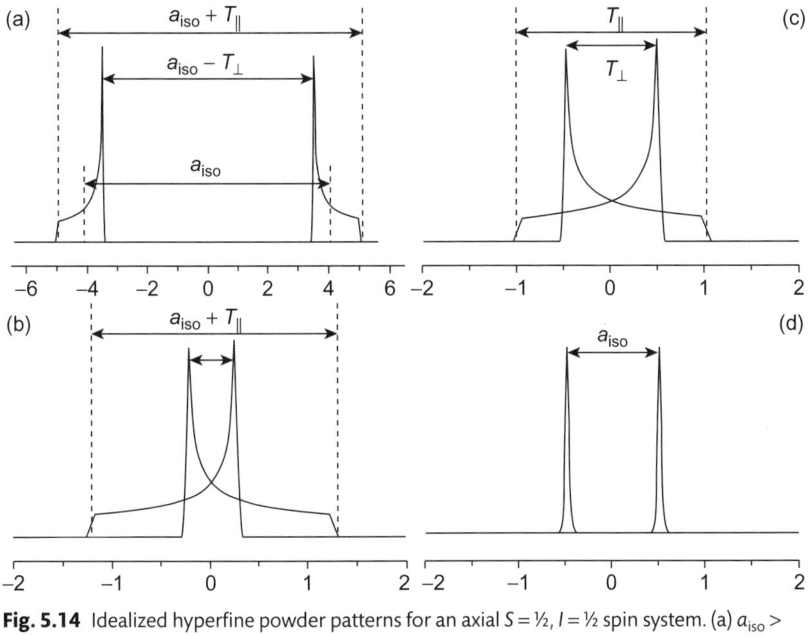

Fig. 5.14 Idealized hyperfine powder patterns for an axial $S = ½$, $I = ½$ spin system. (a) $a_{iso} > T_{\|}$, (b) $a_{iso} < T_{\|}$, (c) $a_{iso} = 0$, $T_{\|} \neq 0$, and (d) $a_{iso} \neq 0$, $T_{\|} = 0$. The transitions may be observed by EPR (X axis in *field units*, where zero represents the resonant position of the electron Zeeman energy) or in a hyperfine ENDOR spectrum (X axis in *frequency units*, where zero represents the resonant frequency of the nuclear Zeeman energy).

with respect to $\boldsymbol{B}$ (expressed through variations in θ). The resulting profile of the hyperfine spectrum will depend on the relative magnitudes of a_{iso} and **T**, as expressed in eqns 5.26 and 5.28.

To illustrate this, consider the possible hyperfine spectra for a two-spin axial system ($S = ½$, $I = ½$) possessing an isotropic g value, and characterized by the conditions where a) $a_{iso} > T_{\|}$, b) $a_{iso} < T_{\|}$, c) $a_{iso} = 0$, $T_{\|} \neq 0$, and d) $a_{iso} \neq 0$, $T_{\|} = 0$. The resulting absorption profiles are shown in Fig. 5.14. The X axis could be specified in field units (as in an EPR spectrum) or frequency units (as in an ENDOR spectrum, see Chapter 9). In the first case, assuming an axial hyperfine ($A_{xx} = A_{yy} = A_{\perp}$; $A_{zz} = A_{\|}$) then owing to the large a_{iso} value, the two hyperfine envelopes are well separated from each other (Fig. 5.14a). The two separate envelopes or subspectra arise since $m_I = \pm½$. The outer edges of the envelope arise when $\theta = 0°$, and owing to the probability the intensity of the inner peaks increases as θ approaches 90° (analogous to the variation in peak intensities for an $S = ½$, $I = 0$ spin system). The resulting a_{iso} and T components of the hyperfine can be readily extracted from the spectrum.

In the second case, a_{iso} is very small relative to $T_{\|}$, so the two hyperfine envelopes now overlap. Once again, the outer peaks arise from $a_{iso} + T_{\|}$ ($\theta = 0°$). At this orientation, the principal axis of the hyperfine interaction is parallel to $\boldsymbol{B}$, so that $\boldsymbol{B}_{local}$ is reinforced between μ_{elec} and μ_{nuc} leading to the larger $T_{\|}$ value (Fig. 5.14b). The inner peaks arise from $a_{iso} - T_{\perp}$ ($\theta = 90°$). In the third case, a_{iso} is zero since the hyperfine is purely dipolar (Fig. 5.14c). It should be noted, that this particular shape is a general example occurring for any orientationally dependent dipolar coupling between two spins and is common in solid state NMR (sometimes referred to as a *Pake doublet*). Finally, in the last example (Fig. 5.14d), the hyperfine is purely isotropic, so only a pair of lines will appear in the spectrum without any dipolar contribution to broaden and change the profile of the pattern.

Fig. 5.14 shows the idealized absorption lineshapes for a simple two-spin axial system with an isotropic g value. However, the complexity of the powder lineshapes grows as the g anisotropy increases, as the number of interacting nuclei increases ($n > 1$), as the nuclear spin I increases ($I > ½$), and even as the number of electron spins increases ($S > ½$). Multi-frequency EPR and hyperfine measurements, combined with computer simulations, are vital in such cases in order to fully analyse the experimental spectra.

5.5 The anisotropic quadrupole P interaction

In real systems, the nuclear spin can be aligned by a number of factors, including the local field generated by the electron $\boldsymbol{B}_{local}$, the external field $\boldsymbol{B}$, and the local electric field gradient. The direction of the nuclear spin angular momentum is dependent on the shape of the nucleus. Therefore when $I > ½$, the electric field gradient can align the charge and thus the nuclear spin.

These gradients are caused by the electron distribution in the surroundings so that the *electric quadrupole moment* (Q) of the nucleus will interact strongly with the electric field gradients. Analysis of the resulting *nuclear-quadrupole parameter matrix* (**P**) therefore affords valuable information on the nature of the atomic orbitals at the nucleus and the electron distributions.

The quadrupole interaction, when smaller than the nuclear Zeeman and hyperfine interactions, only shifts the energy levels according to the nuclear states m_I.

The quadrupole energy (or energy of alignment) is thus represented by the spin-Hamiltonian:

$$\hat{H}_Q = \hat{\boldsymbol{I}}^T \cdot \mathbf{P} \cdot \hat{\boldsymbol{I}} \tag{5.31}$$

Here **P** is a tensor which is traceless and symmetric as defined by:

$$\mathbf{P} = P\begin{bmatrix} \eta-1 & 0 & 0 \\ & -\eta-1 & 0 \\ & & 2 \end{bmatrix} \tag{5.32}$$

It is important to emphasize that **P** is a symmetric and traceless tensor.

and where P is given as:

$$P = \frac{e^2qQ/h}{4I(2I-1)} \tag{5.33}$$

The principal values of the nuclear quadrupole tensor are $P_{xx} = [-(e^2qQ/(4h))](1-\eta)$, $P_{yy} = [-(e^2qQ/(4h))](1+\eta)$, and $P_{zz} = e^2qQ/(2h)$. The asymmetry parameter (η) is equal to $(P_{xx}-P_{yy})/P_{zz}$ with $|P_{zz}| > |P_{yy}| > |P_{xx}|$. Since the quadrupole tensor **P** is traceless, it is determined, apart from its orientation, by only two parameters. Usually in the literature, the two quantities e^2qQ/h and η are reported. The dimensionless *asymmetry parameter* η also provides a measure of the field gradient deviation from uniaxial symmetry ($\eta = 0$ for uniaxial symmetry).

Since the electron spin does not appear in eqns 5.31–5.33, the quadrupole coupling does not contribute to the frequencies of the EPR transitions (to first order). In other words, since the selection rules for EPR transitions are $\Delta m_S = \pm 1$ and $\Delta m_I = 0$, quadrupole effects to first order are usually not observed in EPR spectra but are readily observed in hyperfine spectra (such as ENDOR, see Chapter 9) where the selection rules are $\Delta m_S = 0$ and $\Delta m_I = \pm 1$.

5.6 Combinations of g and A anisotropy

At the start of this chapter, the CW EPR spectra for a vanadyl ion in different phases are shown. Although the next chapter is devoted to transition metal ions and inorganic radicals, it is instructive to finish this chapter by summarizing how the combined **g** and **A** anisotropy accounts for the profile of the solid state VO^{2+} spectra shown earlier in Fig. 5.1.

The vanadyl ion (VO^{2+}) possesses a d^1 electron configuration and therefore behaves as a two-spin system with $S = ½$ and $I = 7/2$. Owing to the square pyramidal structure of the ion, an axial **g** and **A** tensor is expected in the solid state, whereas an isotropic spectrum is observed in fluid solution. In the particular example shown in Fig. 5.15, the experimental g values have the form $g_{\|} < g_{\perp} < g_e$ while the experimental hyperfine values have the order $A_{\|} > A_{\perp}$; the reasoning for this is explored in more detail in Chapter 6. In the isotropic spectrum, the inhomogeneously broadened lines arise due to incomplete averaging of the components of the **g** and **A** tensors (see Chapter 8). In particular, the differing field separations of the parallel and perpendicular components of each m_I transition will result in some lines having narrower width compared to others in the fluid solution spectrum (hence the unequal distribution of peak heights and widths). If the vanadyl ion was tumbling fast enough, eight lines of equal intensity and width would be observed (section 8.3).

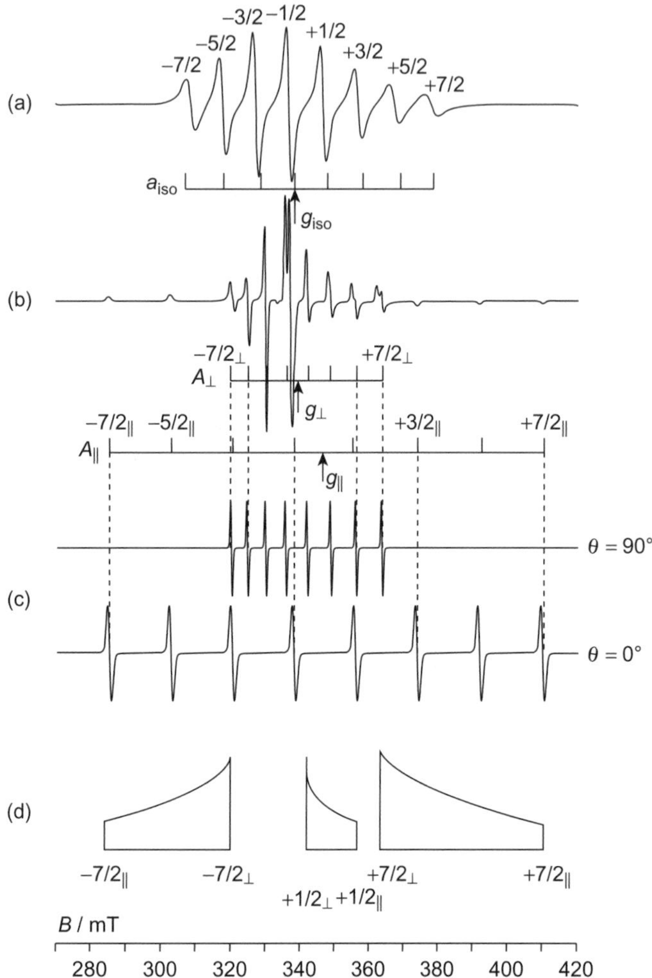

Fig. 5.15 CW EPR spectra of a VO^{2+} ion in (a) the isotropic, and (b) powder (frozen solution) phase. In (c), the two extreme orientations in this axial case (θ = 0° and 90°) extracted from the single crystal spectra are shown in derivative mode. Finally in (d), three (out of the possible eight) absorption traces for the individual m_I transitions −7⁄2, +1⁄2, and +7⁄2 are shown to reveal how the quasi axial **g** anisotropy is manifested in these individual m_I components.

The spin Hamiltonian parameters of the VO^{2+} ion in Fig. 5.15 were: *isotropic* $g_{iso} = 1.971$, $a_{iso} = 9.75$ mT; *anisotropic* $g_{xx} = 1.982$, $g_{yy} = 1.980$, $g_{zz} = 1.952$, $A_{xx} = 5.7$, $A_{yy} = 6.1$, $A_{zz} = 17.3$ mT. Although the **g** and **A** anisotropy is rhombic, at 9.5 GHz the spectrum appears quasi-axial.

In the powder spectrum (Fig. 5.15b) all orientations are simultaneously present and contribute to the powder pattern, as discussed in section 5.3. This spectrum is composed of the eight individual powder patterns due to the $2I + 1$ number of allowed $\Delta m_I = 0$ EPR transitions. As each is influenced by the large **g** and **A** anisotropy, the outer traces are influenced to a great extent. Three of the individual m_I traces revealing the axial profile of the **g** tensor are shown in Fig. 5.15(d) (parallel and perpendicular components of −7⁄2, +1⁄2, and +7⁄2 transition). As discussed in section 5.3, owing to the orientational dependence of the **g** tensor (eqn 5.13), as the crystal is rotated with respect to the applied field, the resonance field will also change. As **A** is similarly dependent on crystal orientation, through the dipolar term, the observed hyperfine coupling will change.

The deconstructed profiles in Fig. 5.15 therefore illustrate how the powder spectrum can be analysed as a summation of the single crystal spectra, with anisotropy in both **g** and **A** determining the final shape of the spectra.

5.7 Summary

- The interaction of an electron spin with the applied field (***B***) is orientationally dependent.
- In the solid state, spin-orbit coupling is the source of the anisotropic **g** tensor.
- The angular dependency terms are defined with respect to $g(\theta, \phi)$, setting limits on the symmetry and magnitude of g.
- The principal directions of **g** with respect to the molecular axes (i.e. the diagonal g_{xx}, g_{yy}, g_{zz} values), cannot be determined from the powder spectrum.
- The intensity of a powder EPR pattern depends on the sum of contributions for each molecular orientation within the unit sphere.
- The interaction of an electron spin with a nuclear spin is orientationally dependent.
- In the solid state, the hyperfine interaction is defined with an anisotropic **A** tensor.
- The anisotropic hyperfine **A** tensor is composed of an isotropic (a_{iso}) and pure dipolar (**T**) component.
- Hyperfine powder lineshapes are dependent on the relative magnitudes of a_{iso} and **T**.
- The anisotropic nuclear quadrupolar **P** interaction, for $I > ½$ nuclei, can be observed in hyperfine spectra but not in EPR spectra.

5.8 Exercises

5.1) Sketch the expected first derivative profiles of the resultant powder EPR spectra for the idealized hyperfine patterns shown in Fig. 5.14.

5.2) A 2 × 2 matrix, labelled **M**, is given below. Find the determinant (*det*(**M**)), the transpose ($\mathbf{M}^T$) and the inverse ($\mathbf{M}^{-1}$) of the matrix. Briefly explain how the diagonal of the matrix can be determined, and therefore why **g** (and **A** or **D**, see Chapter 8) is expressed as a diagonal matrix.

$$\begin{bmatrix} 4 & 3 \\ 2 & -5 \end{bmatrix}$$

5.3) Using eqn 5.12, derive an expression for g^2 when the single crystal containing a paramagnetic centre is rotated in the YZ and XY plane, given the corresponding direction cosines are ($l_x = 0, l_y = \sin\theta, l_z = \cos\theta$) and ($l_x = \cos\phi, l_y = \sin\phi, l_z = 0$) respectively, and where θ is the angle between **B** and the z axis.

5.4) Using a suitable graphics program, calculate the angular dependency curve and resonant field position for an $S = ½$ spin system with axial symmetry given $g_\perp = 2.000$ and $g_\parallel = 1.950$ (assume $\nu = 9.5$ GHz).

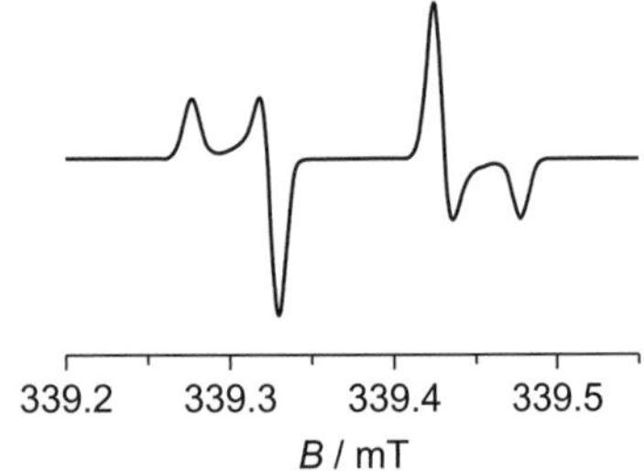

Fig. 5.16 CW EPR spectrum for an $S = I = ½$ spin system with isotropic g and axial **A**.

5.5) The CW powder EPR spectrum for an $S = ½$ spin system with an isotropic g value and a purely axial hyperfine interaction to a proton ($I = ½$), is shown in Fig. 5.16. Determine the axial hyperfine values, in units of MHz, and hence by applying the point-dipole approximation:

$$T = \frac{\mu_0}{4\pi h} g_e \mu_B g_N \mu_N \frac{3\cos^2\theta - 1}{r^3}$$

calculate the electron–proton distance r. What are the limitations of this approach for the determination of r?

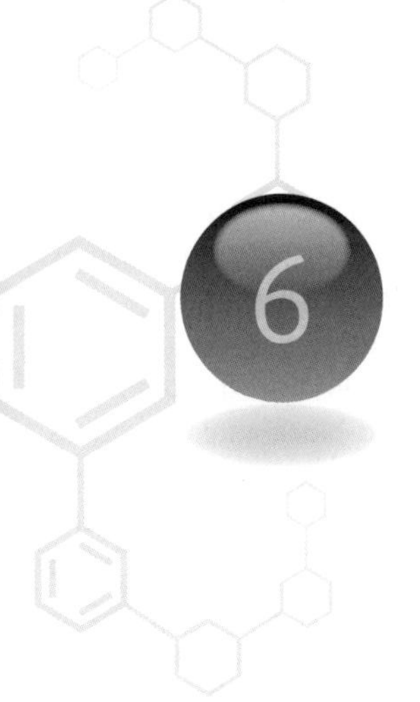

6 Transition metal ions and inorganic radicals

6.1 Introduction

In Chapter 4, the EPR spectra of paramagnetic organic radicals are discussed, where the unpaired electron is primarily associated with sp^x-hybrid orbitals. Many transition metal ions (TMIs), either as isolated dopants in a solid matrix or coordinating an organic ligand in a metal complex, are also paramagnetic. In this case the unpaired electron may be largely confined to metal based d-orbitals. One key difference between the EPR spectra of TMIs compared to organic radicals, is that the former typically exhibit large anisotropic g values which are no longer close to g_e. As we will see, this deviation arises from spin-orbit coupling. Furthermore, the EPR linewidths are often larger compared to organic radicals due to the short relaxation times. In many cases, low temperature measurements are required to lengthen the relaxation time sufficiently for EPR spectra to be observable.

The spectral features depend not only on the nuclear spin of the central ion (Table 6.1) but also on the number and types of ligands surrounding the ion. For example, covalent bonding to oxygen-, nitrogen-, or phosphorus-based ligands can lead to substantial changes in the spin Hamiltonian parameters, so the correct interpretation of the spectra provides detailed information on the interactions with these ligands. For transition metal complexes, the unpaired electron may spend considerable time on the ligands, so the effects of this delocalized spin state must also be taken into consideration in the analysis.

The derivation and interpretation of the **g** and **A** tensors for delocalized spin systems, such as inorganic radicals (i.e. main group ions or molecules containing a ground state with an unpaired electron, but lacking a C–C or C–H bond), is analogous to that involved in analysing the spectra of TMIs. For these radicals significant spin-orbit coupling and electron delocalization over several centres also contribute to the spectra, so their analysis will also be treated in this Chapter.

Table 6.1 Natural abundances and nuclear spins (I) of first row transition metal ions, and common ligand nuclei. A_0 and b_0 are in units of mT (see Appendix A).

Nuc	% Abu.	I	A_0	b_0
^{13}C	1.1	½	134.8	3.8
^{14}N	99.6	1	64.62	1.981
^{17}O	0.04	5/2	−187.8	−6.0
^{19}F	100	½	1887	62.8
^{31}P	100	½	474.8	13.1
^{47}Ti	7.4	5/2	−27.9	−1.1
^{49}Ti	5.4	7/2	−27.9	−1.1
^{51}V	99.8	7/2	148.6	6.2
^{53}Cr	9.5	3/2	−26.7	−1.5
^{55}Mn	100	5/2	179.7	−8.9
^{57}Fe	2.12	½	26.7	1.4
^{59}Co	100	7/2	212.2	12.1
^{61}Ni	1.14	3/2	−89.2	−5.4
^{63}Cu	69.2	3/2	213.9	17.1
^{65}Cu	30.8	3/2	228.9	18.3

6.2 Transition metal ions

The EPR spectra of transition metal ions and their complexes can at first sight appear incredibly complex and varied, even for systems with a simple $S = ½$ ground state. Large **g** and/or **A** anisotropy (see Chapter 5), combined with other effects including superhyperfine interactions and non-coincident **g**/**A** axes (discussed in this chapter), can easily lead to misinterpretation of the spectra. The simplest approach to take in

Table 6.2 Commonly observed electron spin states of first row TMIs.

No. d electrons	Ions	Total spin *S*
1	Ti^{3+}	½
	V^{4+}	½
2	V^{3+}	1
3	V^{2+}	3/2
	Cr^{3+}	3/2
4	Cr^{2+}	2
5	Mn^{2+}	5/2
	Cr^{+}	5/2, ½
	Fe^{3+}	5/2, ½
6	Fe^{2+}	2
7	Fe^{+}	3/2
	Co^{2+}	3/2, ½
8	Ni^{2+}	1
9	Ni^{+}	½
	Cu^{2+}	½

Note, for a general introduction to *ligand field theory*, *group theory* and *perturbation theory* used to predict the **g** and **A** tensors, see recommended textbooks in the Bibliography.

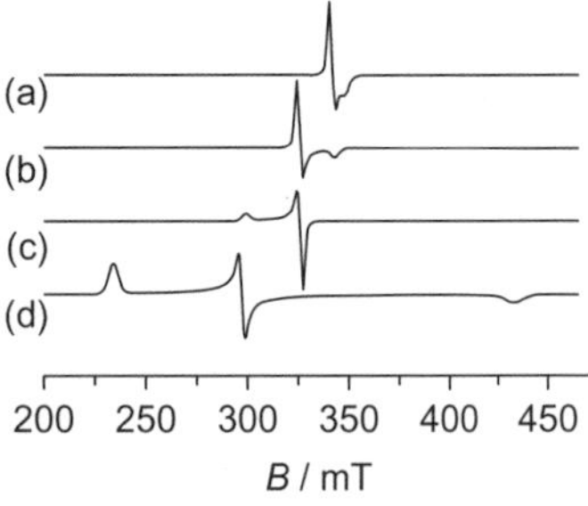

Fig. 6.1 Typical CW EPR spectra for 1st row TMIs with low natural abundances of spin active nuclei ($I \neq 0$); (a) Ti^{3+}, (b) LS Cr^{+}, (c) Ni^{+}, (d) LS Fe^{3+}.

the analysis is to initially extract the experimental g and A values from the spectrum and to rationalize what information these parameters provide about the structure, symmetry or electronic configuration of the central metal ion.

To understand the relationship between the **g** and **A** tensors to the structure, symmetry or electronic properties of the ion, it is useful to predict the expected values for a given system. Derivation of the EPR g and A values for d-block systems requires a good understanding of group theory, which will not be treated here. Nevertheless a qualitative explanation for the origins of these values can be illustrated using ligand field theory for simple d^9 (such as Cu^{2+}, Ni^{+}), d^1 (such as Mo^{5+}, Ti^{3+}, and V^{4+}/VO^{2+}), and d^5 (such as Fe^{3+}, Mn^{2+}) ions (Table 6.2). The combined role of spin-orbit coupling (λ) and ligand field splitting (Δ_{LF}) on the profile of the resulting EPR spectra, are presented for the simplest $S = ½$ cases. High spin systems ($S > ½$) are covered in Chapter 7. The combination of λ and ligand splitting is important since it removes orbital degeneracy from the energy levels.

Some examples of typical CW X-band frozen solution (powder) EPR spectra for TMIs with $S = ½$ spin ground states, including Ti^{3+}, Ni^{+}, Cr^{+} and low spin (LS) Fe^{3+} ions, are shown in Fig. 6.1. Common examples for other TMIs which are dominated by strong metal hyperfine interactions include Mn^{2+}, V^{4+}/VO^{2+}, Cu^{2+}, and LS Co^{2+}, shown in Fig. 6.2. Variation in the coordination environment and symmetry of the ion will then lead to changes in the EPR spectra, as explained in the following sections.

The influence of ligand field splitting on the g tensor

The electronic and magnetic properties (including the g and A values) of d-block elements are dependent on the type and magnitude of the d-orbital splittings. Owing to the direction of the five d-orbitals (d_{xy}, d_{xz}, d_{yz}, d_{z^2}, $d_{x^2-y^2}$), the ligands surrounding the central metal ion cause these orbitals to split into groups with different energies. The magnitude of the splittings will depend on the type of ligands and their arrangement around the central metal ion.

Consider a first row transition metal ion with six ligands (L) placed on the Cartesian axes forming an octahedron (O_h point group symmetry, Fig. 6.3). An electrostatic field (or ligand field) is created by these ligands which destabilizes the orbitals. In the octahedral environment, the energy of the e_g set is raised compared to the t_{2g} set since the d_{z^2} and $d_{x^2-y^2}$ orbitals point directly at the ligands, whilst the remaining orbitals point between these ligands. The two sets of orbitals have an energy separation labelled here as Δ_{LF}. As the strength of the ligand field varies, Δ_{LF} will also vary.

The orbital degeneracy is further lifted for $3d^9$ metal complexes (such as Cu^{2+}) due to a Jahn-Teller distortion. This leads to a tetragonal elongation along the z axis and a corresponding lowering of the symmetry (from O_h to D_{4h}). The square planar arrangement (also of D_{4h} symmetry) can then be formed by removal of the two axial ligands, resulting in considerable stabilization of the d_{z^2} orbital. In both of these D_{4h} cases (Fig. 6.3), the energy levels are singly degenerate (i.e. with symmetry labels *A* or *B*) or doubly degenerate (i.e. with symmetry label *E*).

Next, consider the effects of spin-orbit coupling on the electron in one of the non-degenerate orbitals (Fig. 6.3). The influence of an external magnetic field creates a Zeeman splitting of the non-degenerate orbital, so the g values will arise from spin angular momentum. As a result the g values are expected to be close to g_e. However, suitable excited states can mix with the ground state in a magnetic field such that an admixture of the orbital angular momentum with the spin angular momentum may occur (eqns 5.17, 5.18). In this way, some orbital angular momentum appears in the ground state. The g values will then deviate from g_e and their value will reflect not only the magnitude of the spin-orbit coupling, but also the relative differences in energies between the E_0 and E_n states, depending on the crystal field splitting parameters (e.g. Δ_{LF}) which may be measured by UV-vis spectroscopy.

Notably, if λ is large and the energy difference between the singly occupied molecular orbital, SOMO (E_0), and a low lying excited state (E_n) is small, as occurs in transition metal ions, then a large deviation from g_e will be expected. For organic radicals, the E_n - E_0 energy difference is usually large, so the free spin deviation is correspondingly small ($g \approx g_e$).

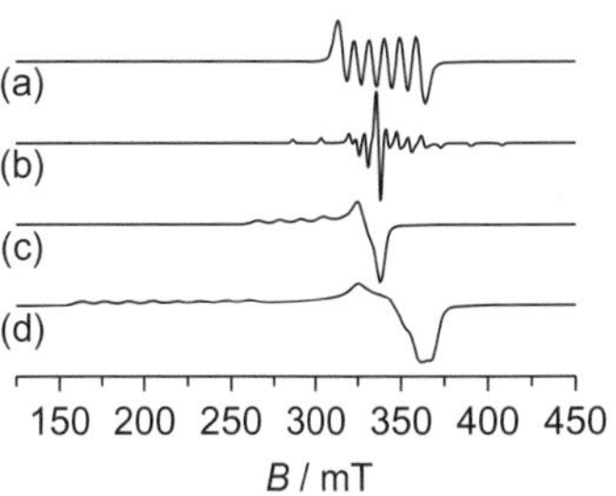

Fig. 6.2 Typical CW EPR spectra for first row TMIs with high natural abundances of spin active nuclei ($I \neq 0$); (a) Mn^{2+}, (b) VO^{2+}, (c) Cu^{2+}, (d) LS Co^{2+}.

The d^9 case

To illustrate the influence of crystal field splitting on the anisotropic g values, consider the case of a 3d^9 ion, such as Cu^{2+}. Assume the ion exists in octahedral symmetry (O_h) and after undergoing a tetragonal distortion produces a system with D_{4h} symmetry, either by elongation or compression along the z axis (Fig. 6.5). In the former case, the unpaired electron resides in the $d_{x^2-y^2}$ orbital. The missing electron in this case can be treated as a positive 'hole', and accordingly this orbital ground state is well separated in energy from other orbitals. Spin-orbit coupling causes the unpaired 'hole' to admix with orbitals, other than $d_{x^2-y^2}$, reintroducing orbital angular momentum.

Crucially, the admixture of d-orbitals depends on the symmetry of the coupled states. When the applied magnetic field is orthogonal to either the x, y, or z axes, the electron is effectively promoted from one orbital to another. This magnetic field induced motion between suitable states (of the correct symmetry) is equivalent to a rotation of the electron about that axis (symmetry representations of R_x, R_y, and R_z). In D_{4h} symmetry, rotations about the z axis belong to the irreducible representation A_{2g},

For simplicity, the elements of g_{ij} from eqns 5.17–5.18 are summarized as:

$$g_{ij} = 2.0023 \pm \frac{n\lambda}{E_0 - E_n}$$

When the number of d electrons is < d^5, then $\lambda > 0$, and when the number is > d^5, $\lambda < 0$. With E_0 - E_n as negative, the -ve sign occurs for half-filled MOs. Crucially, spin-orbit coupling to empty molecular orbitals will produce a negative contribution to g_{ij}.

The value of n depends on which orbitals couple (Fig. 6.4).

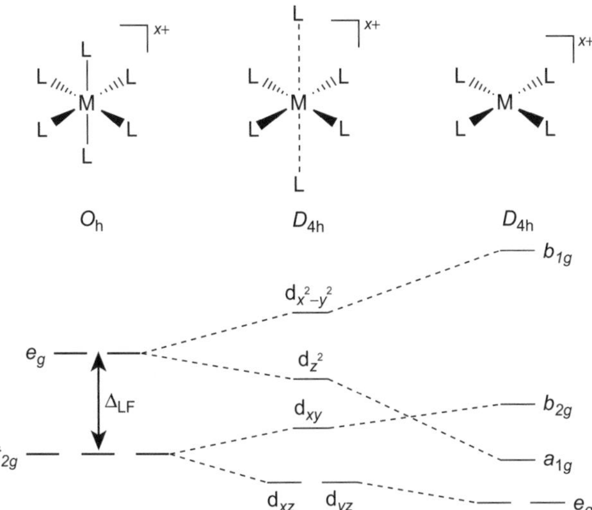

Fig. 6.3 Splitting of the d-orbitals in an octahedral ligand field (O_h point group symmetry), caused by a Jahn-Teller distortion. The subsequent changes after a tetragonal distortion (D_{4h} symmetry caused by elongation along the z axis), and final removal of the ligands to produce a square planar arrangement (D_{4h} symmetry) are shown.

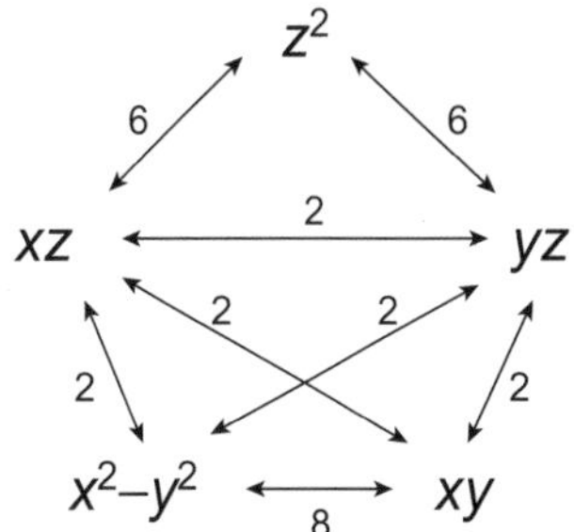

Fig. 6.4 Numerical values of n and corresponding d-orbitals.

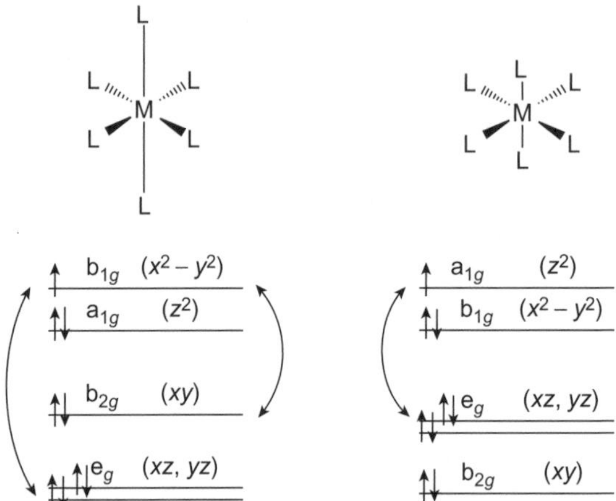

Fig. 6.5 Splitting of the d-orbitals for D_{4h} symmetry, in *elongated* (left) and *compressed* (right) arrangements for a d^9 ion. The orbitals which commute or mix with each other, under the influence of an applied field, are shown by the curved arrows. Crucially this field induced mixing introduces orbital momentum and therefore creates shifts in the g values.

In group theory, if the operator of A_{2g} symmetry transforms an orbital of B_{1g} symmetry (e.g. $d_{x^2-y^2}$), it gives B_{2g} symmetry. This can then couple with other orbitals of B_{2g} symmetry (e.g. d_{xy}). This effectively induces the promotion of an electron between $d_{x^2-y^2}$ and d_{xy} in the magnetic field.

while rotations about the x and y axes belong to the irreducible representation E_g (see Tables B.1 and B.2 in Appendix B). The $d_{x^2-y^2}$ orbital transforms as B_{1g}. Coupling of B_{1g} and A_{2g} gives B_{2g} symmetry (corresponding to d_{xy}). Therefore when the field lies along the z axis, $d_{x^2-y^2}$ can mix with d_{xy} and the magnitude of the resulting g value, with the coefficient $n = 8$ (Fig. 6.4), is given by:

$$g_{zz}(g_{\parallel}) = g_e + \frac{8\lambda}{E_{(x^2-y^2)} - E_{xy}} \tag{6.1}$$

Secondly, when the magnetic field is parallel to the xy plane (R_x, R_y), this will transform as E_g. In this case, coupling of B_{1g} and E_g gives E_g symmetry, corresponding to the degenerate d_{xz}, d_{yz} orbitals. Therefore when the field lies along the x or y axis, $d_{x^2-y^2}$ can mix with d_{xz} or d_{yz} and the magnitude of the resulting g value (in this case $g_{xx,}\, g_{yy} = g_{\perp}$) is given by:

$$g_{xx,yy}(g_{\perp}) = g_e + \frac{2\lambda}{E_{x^2-y^2} - E_{yz,xz}} \tag{6.2}$$

where $n = 2$ in this case (Fig. 6.4). Owing to the difference in energies of the E_0–E_n states in eqns 6.1 and 6.2, anisotropic g values are produced (such that $g_{\parallel} > g_{\perp} > g_e$) and both values are larger than g_e since we are dealing with a d^9 ion in D_{4h} symmetry with elongation along the z axis (Fig. 6.5).

As the symmetry of the ion changes, the ligand field splitting Δ_{LF} will also change, and accordingly different anisotropic g values, which may be axial or rhombic for low symmetry cases, may be expected. For example, if the ion in D_{4h} symmetry experiences compression along the z axis (Fig. 6.5), then according to the crystal field splitting arrangements, the unpaired electron will now occupy the d_{z^2} orbital. The d_{z^2} orbital transforms as A_{1g} (see Table B.1 in Appendix B). When the applied field is aligned along the z axis, this orbital will not mix with any other orbital by spin-orbit coupling. In group theory, the product $A_{1g} \times A_{2g} = A_{2g}$ so that $g_{zz} \approx g_e$. However, when the magnetic field is parallel to the x or y plane (R_x, R_y), the

d_{z^2} orbital (A_{1g}) will couple with the E_g state, producing an E_g state and the $g_{xx,yy}$ value becomes (where $n = 6$):

$$g_{xx,yy}(g_\perp) = g_e + \frac{6\lambda}{E_{z^2} - E_{yz,xz}} \quad (6.3)$$

Once again, different anisotropic g values are expected (for D_{4h} symmetry with compressed tetragonal distortion) with the predicted order $g_\perp > g_\parallel \approx g_e$. In other words, as the energy difference of the E_0–E_n states and the symmetry change, the g values will also change (as illustrated in Fig. 6.6 for Cu^{2+} in different symmetries).

The anisotropic hyperfine A values (see section 5.4) are also dependent on the symmetry and crystal field. Unlike the above g values, the derivation of the A values for d-block ions is not so straightforward. In simple terms, the magnitude of the hyperfine interaction is dependent on the SOMO, and the extent of electron localization on the nucleus in question. The theoretical isotropic (A_0) and uniaxial (b_0) anisotropic hyperfine constants for all the elements are known (see Table 6.1 for selected examples). These values represent the unit spin density in s-, p-, or d-orbitals, so by direct comparison of these theoretical values with the experimental isotropic (a_{iso}) or anisotropic (T) values, one can then calculate the unpaired spin density contribution to the SOMO. The A values for the d^9 cases above are given in the Appendix C.

It is worth noting that the trends in the $g_\parallel$ and $A_\parallel$ values for many square planar d^9 systems are very informative on the nature of the coordinated ligands. For example, as shown in Fig. 6.7, a systematic change in $g_\parallel$ and $A_\parallel$ will be expected for a Cu^{2+} ion coordinated by four nitrogen donor ligands compared to two nitrogens plus two oxygens, or four oxygens (e.g. in a Cu-porphyrin, or a Cu-acetylacetonate complex). The upper left region of the graph corresponds to more negatively charged complexes, while the bottom right corresponds to positively charged complexes. This graph, often referred to as the *Peisach-Blumberg* plot, illustrates how diagnostic even a simple interpretation of the EPR spectra can be since the g and A values will change depending on the interacting ligand.

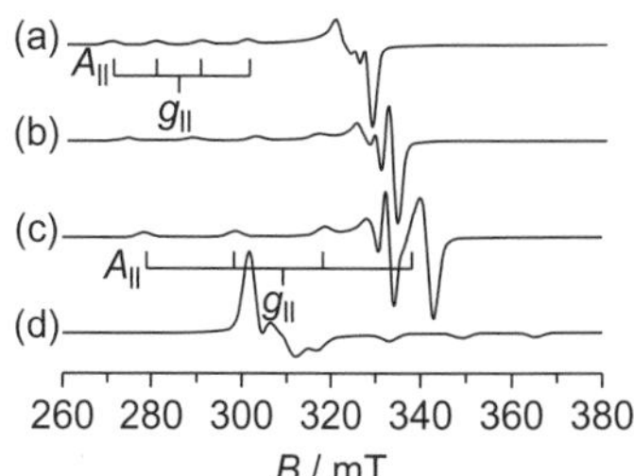

Fig. 6.6 Typical CW EPR spectra of Cu^{2+} in tetragonally distorted crystal fields; (a) elongated octahedral, (b) square pyramidal, (c) square planar, (d) trigonal bipyramidal. Note the systematic changes of $g_\parallel/A_\parallel$ in (a)-(c) and the inverted **g** tensor in (d) due to the d_{z^2} ground state.

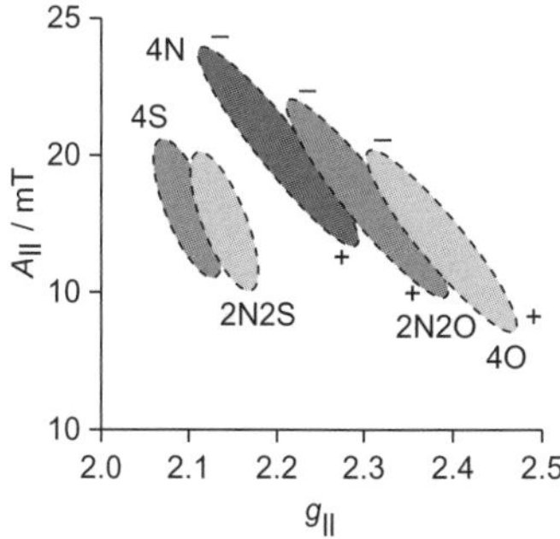

Fig. 6.7 Adapted *Peisach-Blumberg* plot showing the correlation between the $g_\parallel$ and $A_\parallel$ values for square planar $[Cu^{II}(L_4)]$ complexes as a function of the surrounding ligands. Labels + and – refer to positively and negatively charged complexes, respectively.

After J. Peisach and W. E. Blumberg, *Arch. Biochem. Biophys.* 1974, **165**, 691.

The d^1 case

The d^1 configuration (Ti^{3+}, V^{4+}, Cr^{5+}) can be treated using a similar approach to that described above for the d^9 case, assuming the unpaired electron is largely confined to the metal d-orbitals. Typical EPR spectra for Ti^{3+} and VO^{2+} ions are shown in Figs. 6.1 and 6.2, respectively. For Ti^{3+}, the appropriate value of λ must now be included, and since spin-orbit coupling to empty molecular orbitals produces a negative contribution to g, one must carefully consider the effects of the sign in determining the g values.

Consider first the case for a d^1 ion in a distorted tetrahedral environment with D_{2d} symmetry (Fig. 6.8). The distortion lifts the degeneracy of the d_{z^2} and $d_{x^2-y^2}$ orbitals, so the unpaired electron resides in d_{z^2}. Once again the symmetry elements of the orbitals involved and the direction of the applied magnetic field along the z, or x,y directions must be considered. The d_{z^2} orbital will transform as A_{1g}. When the field is parallel to the z axis, this orbital will not mix with any other orbital by spin-orbit coupling (so that $g_{zz} \approx g_e$). When the field is parallel to the x axis, the d_{z^2} orbital (A_{1g}) will couple with the E_g state, producing an E_g state. The derived $g_\perp$ values for the d^1 case in a tetragonally distorted tetrahedral environment are therefore given by:

$$g_{xx,yy}(g_\perp) = g_e - \frac{6\lambda}{E_{z^2} - E_{yz,xz}} \quad (6.4)$$

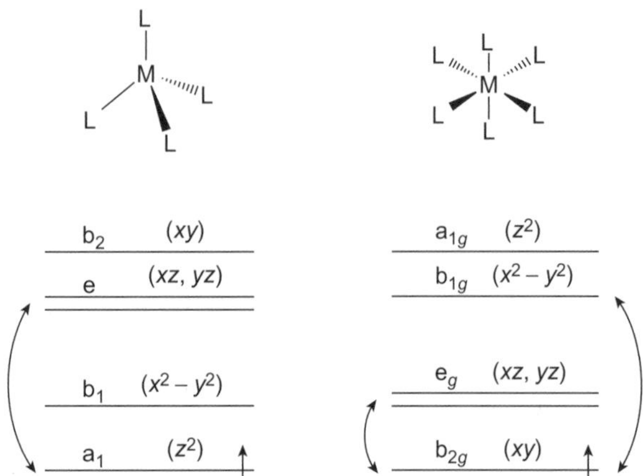

Fig. 6.8 Splitting of the d-orbitals for a distorted tetrahedral (left) and tetragonally compressed octahedral (right) environment for a d^1 ion. The orbitals which mix with each other, under the influence of an applied field, are shown by the curved arrows.

This environment is not commonly encountered for d^1 ions. One of the more frequently studied d^1 ions is VO^{2+}. These ions are close to square planar, so the derivation of the g values can be considered as a tetragonally distorted octahedron with strong axial compression. Simple crystal field arguments predict that the singly occupied orbital should be d_{xy} (Fig. 6.8). The d_{xy} orbital transforms as B_{2g}, so when the field is aligned along the z axis this commutes to B_{1g} (i.e. for $\mathbf{B}$ parallel to z, then $B_{2g} \times A_{2g} = B_{1g}$) and when the field is aligned along the xy axes, this commutes to E_g (i.e. for $\mathbf{B}$ parallel to x or y, then $B_{2g} \times E_g = E_g$). The resulting g values are:

$$g_{zz}(g_{\parallel}) = g_e - \frac{8\lambda}{E_{xy} - E_{x^2-y^2}} \tag{6.5}$$

$$g_{xx,yy}(g_{\perp}) = g_e - \frac{2\lambda}{E_{xy} - E_{yz,xz}} \tag{6.6}$$

With the exception of the d_{z^2} ground state (where $g_{zz} \approx 2.0023$), all principal g values are predicted to be less than 2.0023 with a negative g shift for any other ground state. An example of this for a frozen solution of $[Ti(H_2O)_6]^{3+}$ is shown in Fig. 6.1(a).

The d^5 case

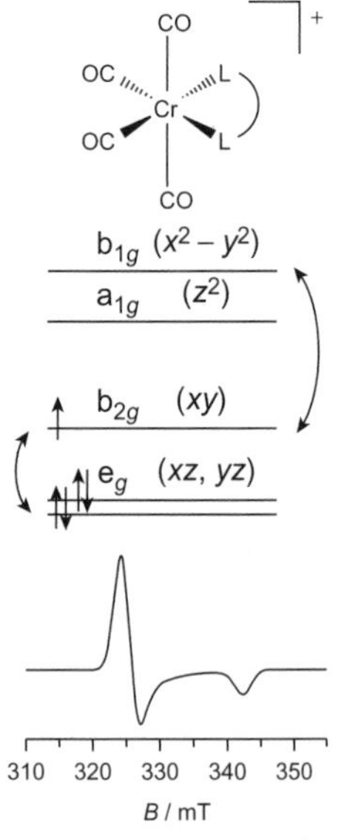

Fig. 6.9 Energy level diagram for a LS d^5 Cr^+ complex ($[Cr(CO)_4L_2]^+$) in a tetragonally distorted octahedral field, and the corresponding EPR spectrum ($g_{\perp} = 2.089 > g_e > g_{\parallel} = 1.983$). L_2 = a bidentate phosphine ligand.

After P. Rieger, *Coord. Chem. Rev.* 1994, **135**, 203.

The EPR spectra of systems with multiple unpaired electrons leading to fine structure are presented in Chapter 7. However, at this point we will make some brief comments on the expected EPR profiles for the low spin d^5 spin systems. The g values and complexity will depend on whether the system is high spin ($S = 5/2$), intermediate spin ($S = 3/2$) or low spin ($S = ½$). For large ligand field splittings, the ion will exist in the low spin state and the g values can be treated in a manner similar to that described above for the d^1 and d^9 cases.

For ions with pure O_h symmetry, the low spin electron configuration may be written as t_{2g}^5 (i.e. all five electrons occupy the t_{2g} set). With tetragonal distortion from O_h to D_{4h} symmetry, the ground state electron configuration becomes either $(d_{xy})^2(d_{xz},d_{yz})^3$ or $(d_{xz},d_{yz})^4(d_{xy})^1$, depending on whether Δ_{LF} is positive or negative, respectively. Consider the six-coordinate $[Cr(CO)_4L_2]^+$ complex as an example (Fig. 6.9). Simple ligand field arguments would predict a $(d_{xz},d_{yz})^4(d_{xy})^1$ ground state, as π-back donation to CO stabilizes d_{xz}, d_{yz} relative to d_{xy}. In this case, d_{xz} and d_{yz} will lie just below the SOMO (i.e. $E_{xz,yz} - E_{xy}$ is small and positive), while $d_{x^2-y^2}$ is empty and much higher in energy (i.e. $E_{x^2-y^2}$ is large and negative). As a result, one would predict that g_{xx} and g_{yy} should be significantly larger than g_e (a positive g shift), due to the admixture of the excited state resulting from promotion of an electron from the doubly occupied $d_{xz,yz}$ to the singly occupied d_{xy} orbital. For g_{zz}, a negative g shift is expected, arising from promotion of the electron from d_{xy} into the empty $d_{x^2-y^2}$ orbital. The resulting g values are predicted to be:

$$g_{xx,yy} = g_e + \frac{2\lambda}{E_{xy} - E_{yz,xz}} \tag{6.7}$$

$$g_{zz} = g_e - \frac{8\lambda}{E_{xy} - E_{x^2-y^2}} \tag{6.8}$$

The $(d_{xy})^2(d_{xz},d_{yz})^3$ configuration is commonly observed for low spin Fe^{3+} systems such as porphyrins, hemes and cytochromes. The g values for this configuration can be obtained by considering only the components of the split t_{2g} orbitals (d_{xy}, d_{xz}, d_{yz}) and using the hole formalism to treat the t_{2g} configuration as a positive hole in the t_{2g} shell. In this manner rhombic g values are predicted, as illustrated in Fig. 6.10. Frequently these spectra are classified according to their ligand field parameters, the tetragonal splitting (Δ/λ) and the rhombicity parameter (V/Δ). Conveniently, the g values can be used to determine the separation of the energies in the t_{2g} set.

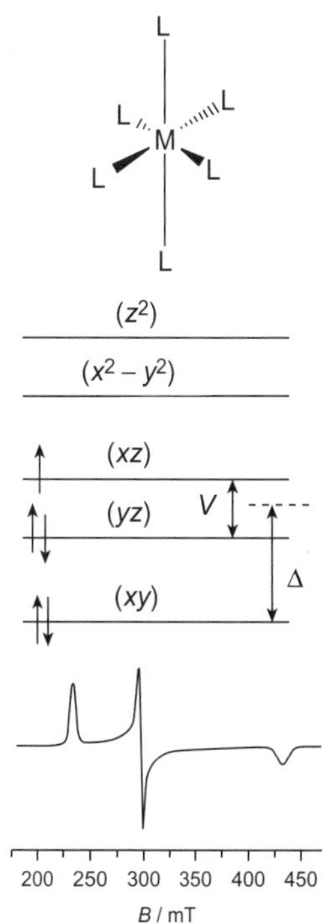

Fig. 6.10 Energy level diagram for a LS d^5 Fe^{3+} ion in a tetragonal field with rhombic distortion (Δ = tetragonal splitting; V = rhombic splitting) and corresponding EPR spectrum.

After P. Rieger, *Coord. Chem. Rev.* 1994, **135**, 203.

In the high spin d^5 state, all orbital angular momentum is quenched due to one electron populating each d-orbital (a half filled atomic shell has zero orbital angular momentum). Thus for purely O_h symmetry, $g \approx g_e$. Some additional fine splitting may be present, caused by dipolar interactions among the five unpaired electrons (see section 7.6). In biological systems, Fe^{3+} and Mn^{2+} usually exist in strongly distorted octahedral environments. As a result, the sixfold degeneracy from the five unpaired electrons splits into three doubly degenerate states called *Kramers doublets* (see Fig. 7.19). If the splitting is much larger than the microwave quantum, and the symmetry is axial, only transitions between the lowest two states ($m_S = \pm½$) are possible. The system therefore behaves as if only two magnetic levels are present and can be described by a spin Hamiltonian with an effective $S = ½$ spin state. However, the effective g value (g_{eff}) is very anisotropic, varying between g = 2–6. Crucially, the large **g** anisotropy observed in these high spin systems arises from the fine structure interactions, unlike the low spin cases where **g** anisotropy arises from the variation in ***B*** with respect to the principal **g** axes.

6.3 Non-coincidence of g and A axes

So far it has been assumed that the **g** and **A** tensors are all diagonal (i.e. with principal values of g_{xx}, g_{yy}, g_{zz}, A_{xx}, A_{yy} and A_{zz}, in which the same principal axis set diagonalizes both matrices). If the paramagnetic centre shares all molecular symmetry elements, then the **g** and **A** matrices are governed by the same rules and will be *coincident*. This occurs for systems with *isotropic*, *axial*, or *rhombic* symmetries (Fig. 6.11a) and the only restrictions on the off-diagonal elements are that $g_{xy} = g_{yz} = g_{xz} = 0$.

However, for *monoclinic* symmetry, only one axis of **g** and **A** coincide with each other (labelled z in Fig. 6.11b), while the remaining two (labelled x,x' and y,y') are

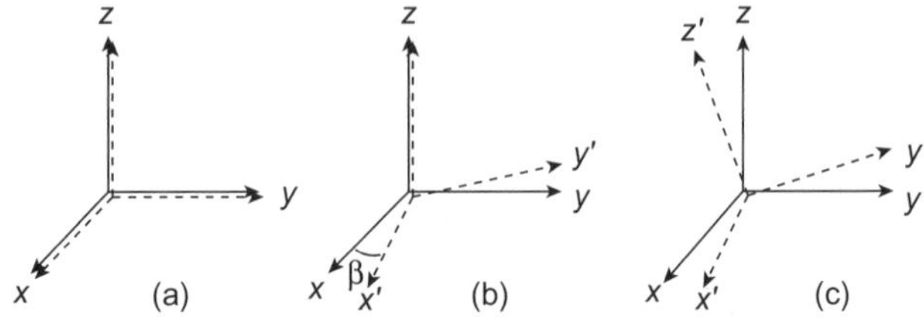

Fig. 6.11 (a) All **g** and **A** axes coincident (*isotropic*, *axial*, and *rhombic* symmetries), (b) one **g** and **A** axis coincident (*monoclinic* symmetry), (c) all **g** and **A** axes are non-coincident (*triclinic* symmetry). In (b) the single angle of non-coincident between *x* and *x′* is defined by β. Solid line = components of **g** tensor, dashed line = components of **A** tensor.

The *rhombic*, *monoclinic*, and *triclinic* terminology used to describe the EPR symmetry should not be confused with analogous terms used in X-ray crystallography.

different. In this case, the restriction on the off diagonal elements is such that $g_{yz} = g_{xz} = 0$. For *triclinic* symmetry, all three principal axes of **g** (*x*, *y*, *z*) and **A** (*x′*, *y′*, *z′*) may be different (Fig. 6.11c). These systems are characterized by their by *non-coincident* **g** and **A** axes.

Whilst the effects of non-coincident **g** and **A** axes may be very small in frozen solution spectra, careful analysis of the spectra is essential for correct interpretation of the data. Analysis usually requires EPR measurements to be performed at multiple frequencies. As a general rule, if two or more matrices have large anisotropies which are of similar magnitude, then the effects of non-coincidence can be readily detected.

Information about non-coincident axes is very useful, since it provides a guide to establish the point symmetries. For example, a system possessing three different sets of *g* and *A* values (g_{xx}, g_{yy}, g_{zz} and A_{xx}, A_{yy}, A_{zz}) but with coincident **g** and **A** axes, must be classified as *rhombic*, so the point symmetry cannot be higher than D_{2h} (Table 6.3) The occurrence of non-coincidence and the extent to which the **g** and **A** axes are rotated from each other depends on how strongly the orbitals mix.

Table 6.3 Relationship between the **g** and **A** tensors, EPR symmetry, and the point symmetry of the paramagnet centre.

EPR symmetry	g and A tensors	Coincidence of tensor axes	Molecular point symmetry
Isotropic	$g_{xx} = g_{yy} = g_{zz}$ $A_{xx} = A_{yy} = A_{zz}$	All coincident	O_h, T_d, O, T_h, T
Axial	$g_{xx} = g_{yy} \neq g_{zz}$ $A_{xx} = A_{yy} \neq A_{zz}$	All coincident	D_{4h}, C_{4v}, D_4, D_{2d}, D_{6h}, C_{6v}, D_6, D_{3h}, D_{3d}, C_{3v}, D_3
Rhombic	$g_{xx} \neq g_{yy} \neq g_{zz}$ $A_{xx} \neq A_{yy} \neq A_{zz}$	All coincident	D_{2h}, C_{2v}, D_2
Monoclinic	$g_{xx} \neq g_{yy} \neq g_{zz}$ $A_{xx} \neq A_{yy} \neq A_{zz}$	One axis of **g** and **A** coincident	C_{2h}, C_s, C_2
Triclinic	$g_{xx} \neq g_{yy} \neq g_{zz}$ $A_{xx} \neq A_{yy} \neq A_{zz}$	Complete non-coincidence	C_1, C_i

In principle the orientation dependence of the axes is lost in randomly oriented samples, since the **g** and **A** matrix principal axes cannot be determined. Nevertheless it is possible to obtain the orientation of a set of matrix axes relative to those of another matrix (but not of course relative to the molecular axes). The first hint that non-coincidence exists is usually that there are too many features in the spectrum or the features appear in the wrong place.

The resonant field positions for the peaks in a randomly oriented powder (see section 5.3) depend on the competition between the extrema in the angle-dependent g and A values. For large values of m_I the resonant field position will be determined largely by the extrema in the effective hyperfine coupling and so appear close to the angles of the hyperfine matrix. By comparison, for small values of m_I, the field positions will be governed primarily by extrema in the effective g value and so appear close to the angles of the **g** matrix. This competition between extrema in **A** or **g** matrices produces a series of features which would normally be equally spaced when **g** and **A** are coincident (to first order) but become more unevenly spaced when **g** and **A** are non-coincident (Fig. 6.12). The separation and field position of the hyperfine lines in the spectra change markedly as the non-coincidence angle β varies.

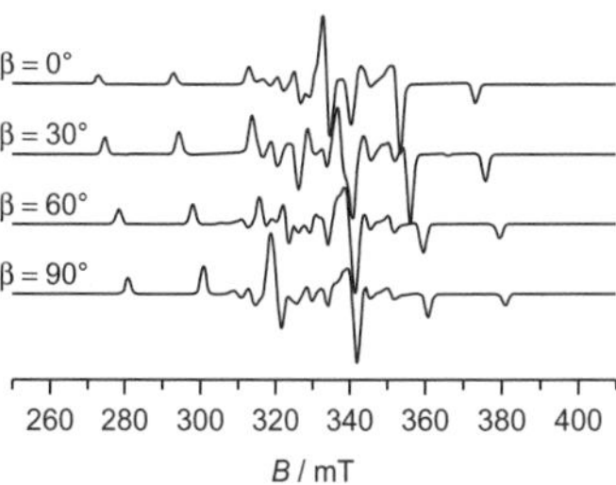

Fig. 6.12 Simulated powder spectra for an $S = \frac{1}{2}, I = \frac{5}{2}$ spin system in monoclinic symmetry with $g_{xx} = 2.10$, $g_{yy} = 2.05$, $g_{zz} = 2.00$, $A_{xx} = 20$, $A_{yy} = A_{zz} = 5.0$ mT for different angles of β (0°, 30°, 60°, and 90°).

6.4 Superhyperfine interaction

If the nuclei surrounding the transition metal ion have non-zero nuclear spin ($I \neq 0$) then a further magnetic interaction occurs with the unpaired electron (called the *superhyperfine* interaction), similar in origin to the interaction between the electron and nucleus of the metal ion itself (the *hyperfine* interaction). The anisotropies in **A** are also observed in the superhyperfine interaction. The resulting superhyperfine splitting will clearly depend on the type and number of interacting nuclei. Two limiting situations arise: i) the ligand superhyperfine splitting is larger than the intrinsic EPR linewidth; or ii) is smaller or of comparable size to the EPR linewidth. In the latter case, advanced EPR techniques are required to resolve these couplings (see Chapter 9).

Note, for organic radicals (Chapter 4), all hyperfine interactions are simply referred to as *hyperfine*.

An example of both cases is shown in Fig. 6.13 for two of the spin systems treated earlier (low spin d^5 Cr^+ and a d^9 Cu^{2+}). In the first case, the unpaired electron interacts

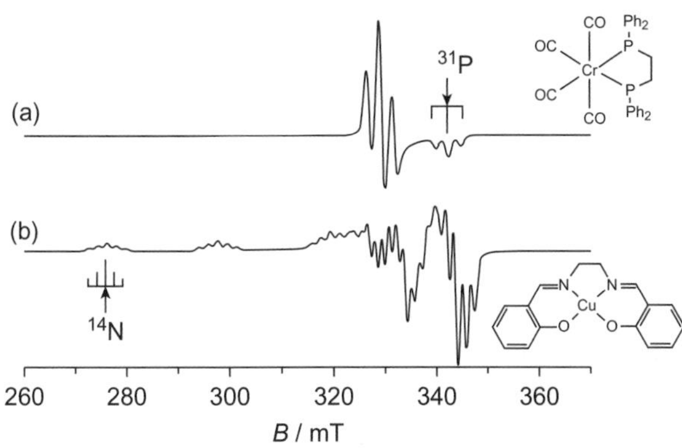

Fig. 6.13 CW EPR spectra of (a) a LS d^5 $[Cr(CO)_4(dppe)]^+$ complex with large ^{31}P superhyperfine splittings and (b) a d^9 $[Cu^{II}(salen)]$ complex with small ^{14}N superhyperfine splittings.

(a) After L. E. McDyre, T. Hamilton, D. M. Murphy, K. J. Cavell, W. F. Gabrielli, M. J. Hanton, and D. M. Smith, *Dalton Trans.* 2010, **39**, 7792, (b) after S. Kita, M. Hashimoto, and M. Iwaizumi, *Inorg. Chem.* 1979, **18**, 3432.

with two equivalent ^{31}P nuclei ($I = ½$). A 1:2:1 triplet pattern is produced according to the $2nI + 1$ rule (see Chapter 4), centred on both $g_{\perp}$ and $g_{\parallel}$. The large and well resolved ^{31}P coupling arises from P 3s-character in the SOMO and partial polarization of inner shell P s-orbitals by spin density on the metal, and through partial P 3p orbital occupation. As the Fermi contact term (A_0) for ^{31}P is very large (Table 6.1), even a small percentage spin density on ^{31}P will produce a relatively large superhyperfine splitting. Although the **g** tensor is anisotropic, the $^{P}\mathbf{A}$ tensor appears isotropic due to this dominant P 3s isotropic contribution.

In the second case, the unpaired electron in a Cu^{2+} complex interacts with two equivalent ^{14}N nuclei. The superhyperfine coupling in this case arises from the ^{14}N orbital contribution to the SOMO and the dominant Cu $d_{x^2-y^2}$ contribution which enhances the overlap with ^{14}N. The A_0 value for ^{14}N is also considerably smaller compared to that for ^{31}P. As a result the anisotropic ^{14}N superhyperfine splittings may be poorly resolved. Unresolved superhyperfine splittings can considerably distort the shape of polycrystalline EPR linewidths, and therefore one must be mindful of their potential contributions when interpreting CW spectra.

6.5 Inorganic radicals

The origin and derivation of the g values for inorganic radicals is very similar to that outlined above for transition metal ions. In this case, the unpaired electron is associated primarily with p-orbitals. This similarity in the derivation of g values will first be demonstrated using O^{-} as an example. A noted feature of inorganic radicals is the considerably greater delocalization of the electron over multiple nuclei, so the additional spin-orbit coupling arising from these nuclei must also be accounted for in the **g** tensor. Once again, the shift in the g values reflects the electronic properties of the radical. Negative Δg shifts indicate promotion of the unpaired electron to an empty orbital, whereas postive Δg shifts suggest promotion from a filled to a half-filled orbital. This will be demonstrated using the diatomic radicals NO and O_2^{-} as examples. Finally, when hyperfine data is available, a great deal of information can be obtained about the structure of the radical, and this will be demonstrated using the triatomic radical NO_2 as an example.

Many inorganic radicals are formed and studied within or stabilized upon a crystal lattice (as opposed to the frozen solution systems discussed above for transition metal complexes). In this case, the local crystal fields can influence the **g** tensor. Some inorganic radicals can therefore be categorized as either *crystal field sensitive* or *crystal field insensitive* species. In the former case, the ground state is usually degenerate and the local crystal field will remove the degeneracy, splitting the two levels proportionally to the intensity of the crystal field environment. Therefore, one must be aware that some components of **g** will change considerably depending on the local field. This occurs for many diatomic (A_2 or AB type) radicals. By comparison, many radicals of the AB_2 type are insensitive to the local field, so in these cases the **g** tensor will not change and in fact will act as a fingerprint in identifying the species.

Monoatomic species

One of the simplest monoatomic inorganic radicals is O^{-}, first observed by EPR in alkali halide single crystals. The radical may exist in an axial or rhombic crystal field, which splits the degeneracy of the three p-orbitals accordingly (Fig. 6.14). The origin of the g values is similar to that described in Chapter 5 (eqns 5.20–5.23) for an unpaired electron in a p-orbital. Based on the electronic structure $2p_x^2 2p_y^2 2p_z^1$, the two predicted g values in an axial crystal field are:

$$g_{xx,yy}(g_\perp) = g_e + \frac{2\lambda}{E_z - E_{x,y}} \quad \text{and} \quad g_{zz}(g_\parallel) \approx g_e \tag{6.9}$$

while the three predicted g values in a rhombic crystal field are:

$$g_{xx} = g_e + \frac{2\lambda}{E_z - E_y},\; g_{yy} = g_e + \frac{2\lambda}{E_z - E_x}, \quad \text{and} \quad g_{zz}(g_\parallel) \approx g_e \tag{6.10}$$

These predicted g values are in very good agreement with the experimental data. For example, the EPR spectrum for a surface stabilized O^- centre is shown in Fig. 6.14, with values of $g_\perp = 2.036 > g_\parallel = 2.0025 \approx g_e$, indicating that the unpaired electron is primarily 2p based within a well-defined axial crystal field environment. When hyperfine information is available, using ^{17}O labelling ($I = 5/2$), then further information on the spin distribution in the radical can be obtained. The $^{17}O^-$ spectrum is also shown in Fig. 6.14, and the resulting experimental ^{17O}A values are $A_\parallel = \pm 10.56 > A_\perp = \pm 1.92$ mT. As described in section 5.4, the **A** tensor can be deconvoluted into isotropic (a_{iso}) and dipolar (**T**) parts according to:

$$\mathbf{A} = a_{iso} I + \mathbf{T} \tag{6.11}$$

When the unpaired electron is confined to a single p-orbital, then an axial hyperfine tensor should be observed. The resulting decomposed ^{17O}A tensor is written as:

$$\begin{bmatrix} +1.92 & & \\ & +1.92 & \\ & & -10.56 \end{bmatrix} = -2.24 + \begin{bmatrix} +4.16 & & \\ & +4.16 & \\ & & -8.32 \end{bmatrix} \tag{6.12}$$

The spin density in the 2s-orbital (labelled ρ^s) can be found using the relation $a_{iso} = \rho^s A_0$ (where A_0 is given in field units, see section 5.4) giving a value of $\rho^s = -2.24/(-187.8) = 0.012$. The $2p_z$-orbital spin density (ρ^p) can likewise be found knowing the theoretical anisotropic dipolar coupling for ^{17}O (Table 6.1) using the analogous relation for the dipolar part, $\rho^p = T/b_0$ (section 5.4) and the experimentally determined value of T (eqn 6.12). However, although the experimentally determined value of ρ^p becomes 0.693, the actual value is 0.82, due to a necessary correction for the negative charge on the coupling caused by the increased nuclear screening which creates a decrease in the anisotropic coupling constant. The combined spin density on O^- determined by EPR is therefore $\rho^{total} = \rho^p + \rho^s = 0.82 + 0.012 = 0.832$.

This anion can occupy sites of C_{3v} symmetry on metal oxide surfaces, such as MgO. As presented in Table 6.3, axial g and A values with coincident tensors are expected for this symmetry, and this is confirmed experimentally. This example illustrates how a simple analysis of the g and A parameters can reveal detailed information on the nature of the radical.

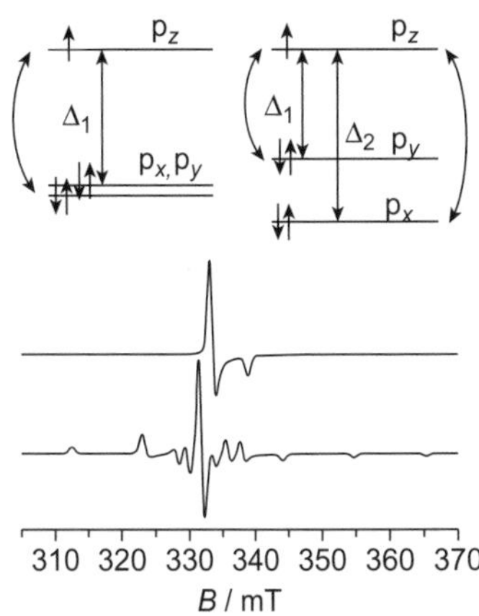

Fig. 6.14 Splitting of the p-orbitals for an O^- radical in an *axial* (left) and *rhombic* (right) crystal field. The CW EPR spectra for unlabelled and ^{17}O labelled O^- in an axial crystal field are shown.

After M. Chiesa, E. Giamello, C. Di Valentin, and G. Pacchioni, *Chem. Phys. Lett.* 2005, **403**, 124.

Diatomic species (π-radicals)

Derivation of the g values for π and π^* radicals is more complex owing to the degenerate ground state expected in these systems. As mentioned earlier, the g values for many diatomic radicals, including the eleven-electron NO, N_2^-, CO^- ($^2\Pi_{1/2}$ state) species or the thirteen-electron O_2^- ($^2\Pi_{3/2}$ state) radicals, are crystal field sensitive, because the

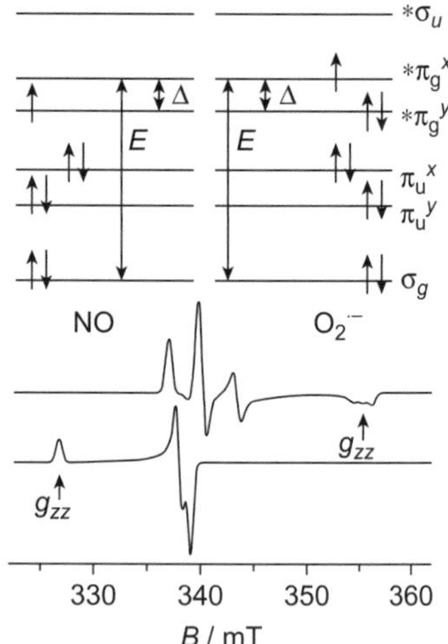

Fig. 6.15 Splitting diagram for the bonding and antibonding orbitals of NO and $O_2^{\cdot-}$. The CW EPR spectra for NO (upper) and $O_2^{\cdot-}$ (lower) are shown. The g_{zz} values will change depending on the magnitude of crystal field splitting Δ.

After M. Chiesa, E. Giamello, and M. Che, *Chem Rev.* 2010, **110**, 1320 and C. Di Valentin, G. Pacchioni, M. Chiesa, E. Giamello, S. Abbet, U. Heiz, *J. Phys. Chem. B* 2002, **106**, 1637.

degeneracy may be lifted by a local field. The predicted g values then depend not only on the orientation of the radical with respect to the applied field, but also on whether the excited state results from promotion of the unpaired electron from a filled to a half-filled orbital or from a half-filled to an empty orbital. This is illustrated in Fig. 6.15, for NO ($g_{xx} = 1.995$, $g_{yy} = 1.998$, $g_{zz} = 1.91$) and $O_2^{\cdot-}$ ($g_{xx} = 2.001$, $g_{yy} = 2.008$, $g_{zz} = 2.077$).

Owing to the electronic configuration of these radicals, a rhombic **g** tensor is predicted. The theoretical expressions for the **g** tensor are rather complex, but the simplified first order treatment which assumes $\lambda < \Delta << E$ (see Fig. 6.15) is given as:

$$g_{zz} = g_e \pm \frac{2\lambda}{\pi_g^x - \pi_g^y},\ g_{yy} = g_e \pm \frac{2\lambda}{\pi_g^x - \sigma_g^z},\ \text{and } g_{xx} \approx g_e \qquad (6.13)$$

The sign of T also depends on the sign of g_N, but can also be influenced by the spin polarization effects.

The z-direction is chosen along the internuclear axis, and it follows that $g_{yy} > g_{xx} >> g_{zz}$ for NO while $g_{zz} >> g_{yy} > g_{xx}$ for $O_2^{\cdot-}$. Whilst the g_{yy} and g_{xx} values are predicted to be close to g_e for both systems, the g_{zz} value will vary depending on the local crystal field splitting since the degeneracy in $2p\pi_g^{x,y}$ ($\Delta = \pi_g^x - \pi_g^y$) is lifted both by spin-orbit coupling and by the local field. For this reason, these radicals produce a heterogeneity of g_{zz} values which provide an excellent probe of local fields, particularly on oxide surfaces.

The hyperfine tensor for these radicals can be treated using eqn 6.11. In this case, the dipolar part of the hyperfine tensor is not exactly axial and is composed of two traceless components ($-T$, $2T$, $-T$ and $2T'$, $-T'$, $-T'$). This situation arises because although the unpaired electron is mostly confined to one of the two π* orbitals, a partial admixture between π_g^x and π_g^y occurs *via* spin-orbit coupling. This enables the $2p_y$ (*via T*) and $2p_x$ (*via T'*) orbital spin densities (ρ^p) to be determined, whilst the 2s spin density (ρ^s) is found knowing the experimental a_{iso} and theoretical A_0 values.

For example, the powder EPR spectrum of the surface adsorbed NO radical is shown in Fig. 6.15. The g values were given above and the corresponding experimental hyperfine values are $A_{xx} = \pm0.257$, $A_{yy} = \pm3.2$, $A_{zz} = \pm0.886$ mT. The absolute signs of these hyperfine couplings (+ or −) cannot be determined from a powder EPR spectrum (see section 5.3); extraction of the correct signs requires analysis of the single crystal spectra. However, the only sensible choice for NO arises when $A_{yy} > 0$, with A_{xx}, $A_{zz} < 0$ so that the resulting a_{iso} is positive. This resulting tensor can then be decomposed to give:

$$\begin{bmatrix} -0.257 & & \\ & +3.2 & \\ & & -0.886 \end{bmatrix} = 0.686 + \begin{bmatrix} -1.36 & & \\ & 2.72 & \\ & & -1.36 \end{bmatrix} + \begin{bmatrix} 0.42 & & \\ & -0.21 & \\ & & -0.21 \end{bmatrix} \qquad (6.14)$$

The sign of a_{iso} depends on the sign of g_N, but also on the extent of s-orbital contribution to the SOMO through the direct or indirect spin polarization mechanism.

The 2s-orbital (ρ^{2s}) spin density is given by $a_{iso} = \rho^s A_0$. With $a_{iso} = 0.686$ mT and $A_0 = 64.6$ mT for ^{14}N (Table 6.1), this gives $\rho^{2s} = 0.686/64.6 = 0.011$. This small spin density comes from the direct contribution of the ^{14}N 2s orbital to the molecular SOMO and the spin polarization of the ns-orbitals. The spin densities in the $2p_{y,x}$-orbitals can be found from the experimental T (1.36 mT) and T' (0.21 mT) values, knowing the theoretical anisotropic dipolar coupling for ^{14}N ($b_0 = 1.98$ mT), and using the expressions $\rho^{p_y} = T/b_0$ and $\rho^{p_x} = T'/b_0$. The values obtained are $\rho^{p_y} = 0.686$ and $\rho^{p_x} = 0.106$, so the total unpaired spin density on the ^{14}N atom in NO is estimated to be $\rho^{p_y} + \rho^{p_x} + \rho^s =$

0.686 + 0.106 + 0.011 = 0.803. A similar approach can be adopted to determine the spin densities of other diatomic π radicals such as $N_2^{\cdot-}$ and $O_2^{\cdot-}$.

Triatomic radicals

In order to appreciate the significance and origin of the g values for triatomic AB_2 type radicals (such as the seventeen-electron $CO_2^{\cdot-}$ or NO_2 species, and the nineteen-electron $SO_2^{\cdot-}$ or ClO_2 species) one must consider the variation in the bonding scheme depending on whether the radical is linear (180°) or bent (e.g. 90°). The correlation diagram for the orbital energies as a function of the bending angle was given by Walsh (summarized in Fig. 6.16). The σ_g, σ_u, and π_u levels are all bonding orbitals, with π_u being doubly degenerate. The π_g level is also doubly degenerate, and since it is confined to the outer atoms, it is considered non-bonding. The $^*\pi_u$, $^*\sigma_g$, and $^*\sigma_u$ levels are all antibonding. Molecules possessing fewer than seventeen electrons can accommodate all the electrons in the bonding and non-bonding orbitals. In this case, the molecule will not bend to relieve any antibonding character and will be linear.

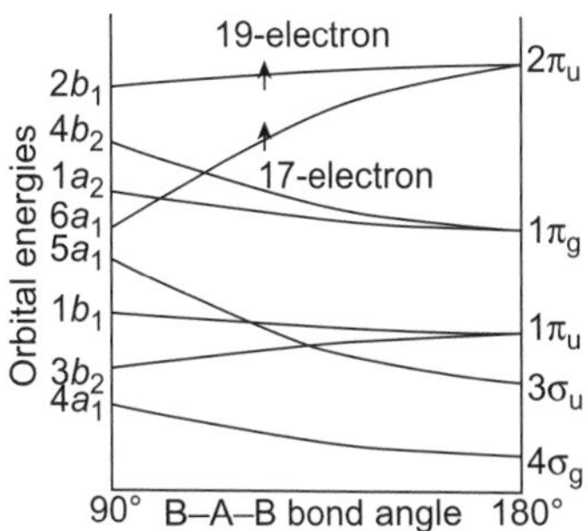

Fig. 6.16 Walsh orbital energy diagram for triatomic molecules (AB_2) showing the variation in B-A-B angle from linear (180°) to bent (90°).

However, for a seventeen-electron radical (such as $CO_2^{\cdot-}$ or NO_2), the unpaired electron must enter the $^*\pi_u$ antibonding orbital (Figs 6.16 and 6.17). As the molecule bends, this $^*\pi_u$ antibonding orbital becomes increasingly built from the s-orbital on the central A atom, until at 90° it occupies an s-lone pair on A. This lowering in energy, brought about by the inclusion of the s-orbital, favours a bent molecule. As we shall see shortly, it is possible to determine this bond angle from the hyperfine tensor. For a nineteen-electron radical ($SO_2^{\cdot-}$, ClO_2, or $O_3^{\cdot-}$), the unpaired electron enters the $2b_1$ orbital. The energy of this orbital is now only slightly dependent on the bond angle (Fig. 6.16), and therefore will have little further effect on the bond angle of these radicals.

The predicted g and A values for these triatomic species can be illustrated using NO_2 as an example. The $3a_1$ orbital is delocalized and built from both s- and p-orbitals on the central atom (nitrogen). As a result the hyperfine tensor is expected to possess a relatively large a_{iso} component associated with ^{14}N, and an anisotropic dipolar contribution (T) possessing a maximum value when NO_2 is aligned with the field along the C_2 axis. This should give a simple axial hyperfine tensor (i.e. $A_{zz} = A_{\|}$ and $A_{xx} = A_{yy} = A_{\perp}$). However, as we have seen for diatomic radicals, a small amount of spin polarization of orthogonal orbitals usually occurs, producing small deviations from this symmetry so the hyperfine is expressed as $a_{iso} + [-T, -T, 2T] + [2T', -T', -T']$.

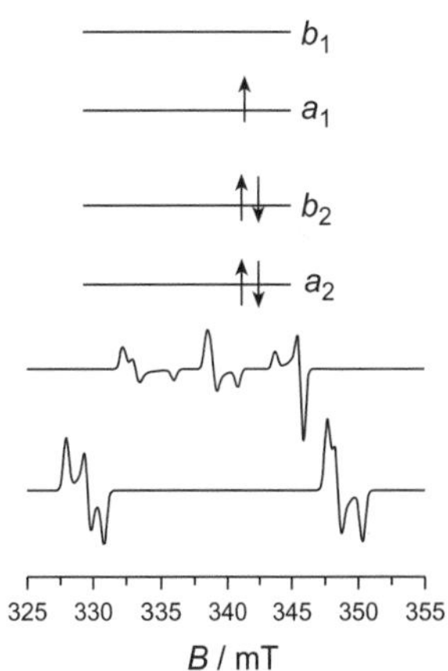

Fig. 6.17 Orbital splitting diagram for a bent seventeen-electron triatomic radical. The CW EPR spectra of NO_2 (upper) and $^{13}CO_2^{\cdot-}$ (lower) are shown.

After M. Chiesa, E. Giamello, and M. Che, *Chem Rev.* 2010, **110**, 1320.

The directions refer to x, y, z where z is the C_2 axis and x is perpendicular to the molecular plane. To a first approximation, a rhombic **g** tensor is predicted depending on the energy separation (Δ) between the levels b_1–a_1, b_2–a_1, and a_2–a_1 (Fig. 6.17) and the combined spin-orbit coupling from N and O. Owing to the magnitude and differences of these energy separations, two g values are predicted to be close to g_e, while the third (g_{yy}) is expected to be much larger and display a negative shift.

These predictions are borne out by the experimental data for NO_2 (i.e. $g_{xx} = 2.005$, $g_{yy} = 1.991$, $g_{zz} = 2.002$, $A_{xx} = \pm 5.27$, $A_{yy} = \pm 4.91$, $A_{zz} = \pm 6.75$ mT). Once again the **A** tensor can be decomposed according to eqn 6.11. The isotropic hyperfine coupling (a_{iso}) for this radical is known to be 5.643 mT, therefore one can assume that all of the A values must be positive. The isotropic and dipolar components of the hyperfine tensor are therefore:

$$\begin{bmatrix} 5.27 & & \\ & 4.91 & \\ & & 6.75 \end{bmatrix} = 5.643 + \begin{bmatrix} -0.373 & & \\ & -0.733 & \\ & & 1.107 \end{bmatrix} \quad (6.15)$$

As discussed above, the dipolar part of the tensor can be decomposed into a larger (T) and a smaller (T') traceless term:

$$\begin{bmatrix} 5.27 & & \\ & 4.91 & \\ & & 6.75 \end{bmatrix} = 5.643 + \begin{bmatrix} -0.613 & & \\ & -0.613 & \\ & & 1.226 \end{bmatrix} + \begin{bmatrix} 0.24 & & \\ & -0.12 & \\ & & -0.12 \end{bmatrix} \quad (6.16)$$

This implies that the unpaired electron is not purely p_z-based, possessing some additional occupancy of the orthogonal p-orbitals. Since the theoretical dipolar hyperfine coupling for ^{14}N 2p is $b_0 = 3.962$ mT, the spin density on the nitrogen 2p-orbitals can be assessed by direct comparison of the experimental dipolar T and T' values with the atomic anisotropic constant of ^{14}N using the formulae:

$$\rho^{2p_z} = c_{p_z}^2 = T/b_0 \quad (6.17a)$$

$$\rho^{2p_z} = c_{p_z}^2 = T'/b_0 \quad (6.17b)$$

The resultant spin densities on the nitrogen $2p_z$ and $2p_x$ orbitals are found to be 0.31 and 0.06, respectively. The fractional occupancy of the ^{14}N s-orbital (ρ^{2s}) may be determined knowing the a_{iso} value and using the theoretical isotropic hyperfine coupling constant (A_0 = 64.6 mT), giving ρ^{2s} = 5.643/64.6 = 0.087. The fraction of the unpaired electron associated with the ^{14}N nucleus is then 0.31 + 0.06 + 0.087 = 0.457. The remaining electron spin density is then shared with the O p_z-orbitals.

Finally, structural information on the geometry of the radical can also be obtained knowing the s ($C_S^2 = 0.087$) and p-orbital ($C_p^2 = 0.37$) occupancies and the hybridization ratio (λ^2) defined as $\lambda^2 = C_p^2 / C_s^2$. For planar triatomic radicals with C_{2v} symmetry, the angle (ϕ) of the radical can be related to λ^2 using:

$$\phi = 2\cos^{-1}\left[\left(\lambda^2 + 2\right)^{-1/2}\right] \quad (6.18)$$

In this case, $\lambda^2 = 0.37/0.087 = 4.253$, giving a value for the angle ϕ of 132° for this bent triatomic species. For tetra-atomic radicals, the two structural angles of ϕ and θ (see Fig. 6.18) can also be extracted from the hybridization ratio according to:

$$\phi = 2\cos^{-1}\left[\frac{1.5}{2\lambda^2 + 2} - \frac{1}{2}\right] \quad (6.19a)$$

$$\cos\phi = \frac{1}{2}\left(3\cos^2\theta - 1\right) \quad (6.19b)$$

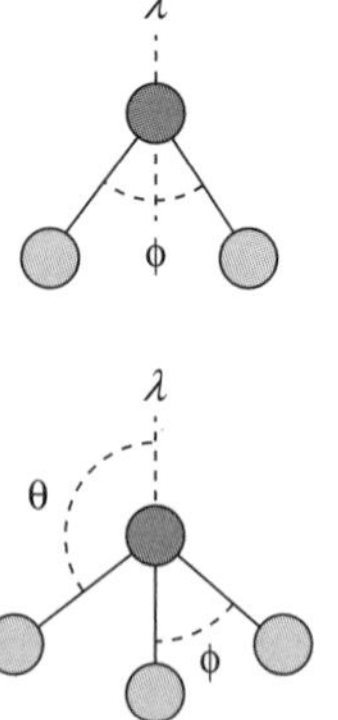

Fig. 6.18 Relationship between the hybridization ratio and the angles for tri- and tetra-atomic radicals.

The above examples illustrate how analysis of the experimental hyperfine coupling provides valuable information not only on the electronic properties but also the structure of simple triatomic radicals (such as NO_2, CO_2^{-}, ClO_2^{-}, SO_2^{-}). A similar approach can be used to extract the structural angles, defined as θ and ϕ, in tetra-atomic radicals (such as PO_3^{2-}, NO_3, CO_3^{-}).

6.6 Summary

- The magnitude of the **g** and **A** tensors of TMIs depend on ligand field splitting (Δ_{LF}) and spin-orbit coupling (λ).

- Shifts of g from g_e depend on the spin–orbit coupling between the SOMO and an empty or half-filled orbital of higher energy.
- Local point group symmetries of TMIs determine the anisotropy in the **g** and **A** tensors.
- Additional features may arise in EPR spectra of TMIs from non-coincident axes and superhyperfine splittings.
- **g** and **A** tensors of delocalized inorganic radicals depend on the crystal field splitting (Δ_{CF}) and spin–orbit coupling (λ).
- The **A** tensor provides information on spin populations and dihedral angles of inorganic radicals.

6.7 Exercises

6.1) The EPR spectrum of a low spin $[Mn^{II}(CN)_5(NO)]^{2-}$ complex in C_{4v} symmetry is characterized by axial g values ($g_{\|} = 1.9892$, $g_{\perp} = 2.0265$). Using this information, account for the likely d-orbital splitting arrangement of the complex and hence explain the magnitude of the g values.

6.2) The frozen solution X-band EPR spectrum (recorded at 9.5 GHz) of a square planar Co(II) complex bearing an axially coordinated substrate, $[Co^{II}(L_2)_2](X)$, is shown in Fig. 6.19. By analysing the spectrum, determine the anisotropic g and A values.

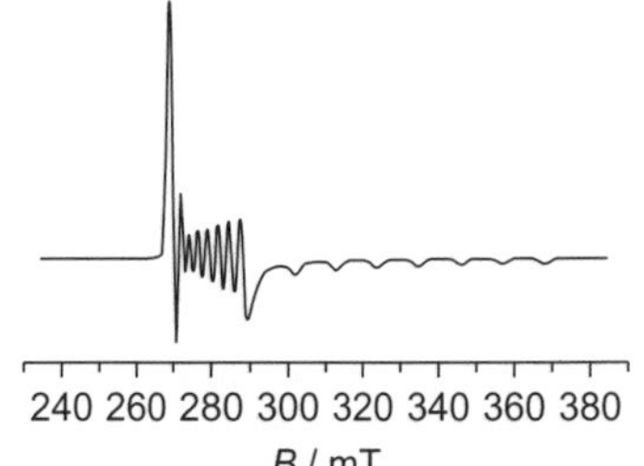

Fig. 6.19 CW EPR spectra of a $[Co^{II}(L_2)_2](X)$ complex ($I = 7/2$ for Co).

6.3) Predict the relative magnitude and anisotropy of the g values for the two d^9 Ni(I) complexes given below, i.e. *trans*-$[Ni^I(L_4)X_2]$ and trigonally distorted *trans*-$[Ni^I(L_4)X_2]$.

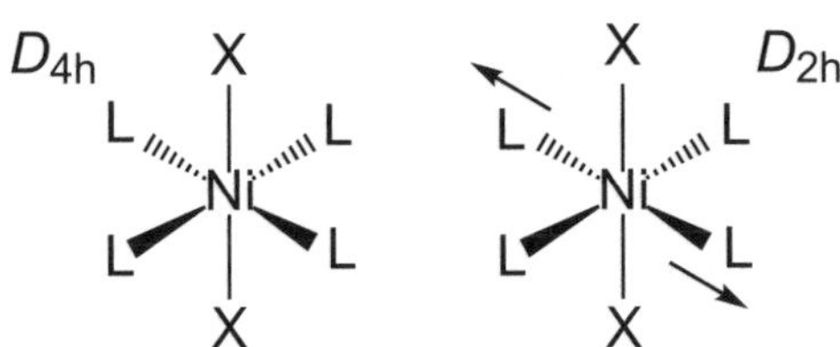

6.4) γ-irradiation of dimethyl glyoxime at low temperature generates an iminoxy radical with the following proposed formulation:

HON=C(CH$_3$)–C(CH$_3$)=N–O•

Assuming the CN•O fragment of this radical can be considered as possessing C_{2v} symmetry, and given the spin Hamiltonian parameters of the radical were determined to be $g_{xx} = 2.0095$, $g_{yy} = 2.0063$, $g_{zz} = 2.0026$, $A_{xx} = \pm 2.5$, $A_{yy} = \pm 2.5$, $A_{zz} = \pm 4.5$ mT, estimate the value of the $C\hat{N}O$ angle (ϕ).

6.5) The anisotropic g and A values for a surface stabilized $^{13}CO_2^-$ radical were determined to be $g_{xx} = 2.0026$, $g_{yy} = 1.9965$, $g_{zz} = 2.0009$, $A_{xx} = \pm 18.1$, $A_{yy} = \pm 17.7$, $A_{zz} = \pm 22.4$ mT. Using this information, estimate the ^{13}C unpaired spin density and orbital occupancy. Comment on the anisotropy of the **g** tensor with respect to the symmetry of the radical.

7 Systems with multiple unpaired electrons

7.1 Introduction

A summary of interactions in EPR (see also Chapter 1):

- Electron spin with magnetic field (electron Zeeman interaction)
- Nuclear spin with magnetic field (nuclear Zeeman interaction)
- Electron spin with nuclear spin (hyperfine interaction)
- Nuclear quadrupole interaction
- Electron spin with electron spin (subject of this chapter!)

The spin–spin interaction between the unpaired electron and nuclei with spin $I \neq 0$ results in splitting of EPR signals into hyperfine patterns. But what about interactions between several unpaired electrons? This is not as commonly observed as the hyperfine interaction, as many paramagnetic compounds have just one unpaired electron. Nonetheless, spin–spin interaction between unpaired electrons affects EPR spectra in a number of important systems, such as:

- liquid or frozen solutions of paramagnetic compounds at high concentration;
- neat paramagnetic liquids or solids (e.g. in the absence of solvent);
- organic molecules with two or more unpaired electrons, e.g. diradicals;
- transition metal ions with high spin (e.g. $S > ½$).

The basic principles of electron spin–electron spin interactions in these systems are somewhat similar to the electron spin–nuclear spin (i.e. hyperfine) interaction; however, the effect on the EPR spectra is rather different. This chapter considers the EPR spectra of systems with multiple unpaired electrons. A brief description of the underlying theory is supplemented by examination of EPR spectra for each type of system.

7.2 Exchange interaction

The exchange interaction is usually considered an isotropic interaction. The anisotropic contribution to exchange is negligible for organic radicals but can make a contribution to radicals containing heavy elements.

As described in Chapter 2, the hyperfine interaction includes (i) an isotropic component (Fermi contact interaction) that results from the overlap of the electron orbital with the nucleus, and (ii) an anisotropic component (dipole–dipole interaction). Similarly, the interaction between two unpaired electrons includes (i) an exchange interaction that is usually isotropic (observed when the orbitals hosting the unpaired electrons partially overlap), and (ii) an anisotropic dipole–dipole interaction. In fluid systems, rapid tumbling averages the anisotropic interaction to zero if it is relatively weak (which may not be the case for some systems with multiple unpaired electrons, e.g. organic triplets, section 7.4, and transition metal ions with $S > ½$, section 7.6), and hence EPR spectra are only affected by the exchange interaction.

Theory of exchange interaction

The exchange interaction is a direct consequence of the Pauli exclusion principle. For a two-electron system, it is described by the following term which adds to the spin Hamiltonian (eqn 7.1):

$$\hat{H} = J \cdot \hat{\mathbf{S}}_1^{\mathrm{T}} \cdot \hat{\mathbf{S}}_2 \tag{7.1}$$

Here $\hat{\mathbf{S}}_1$ and $\hat{\mathbf{S}}_2$ are the electron spin angular momentum operators for the two electrons and J is the exchange coupling constant. The exchange interaction splits the energy of the system into singlet ($S = 0$) and triplet ($S = 1$) states (Fig. 7.1), where J equals the energy difference between the two states. A positive value of J corresponds to the singlet being lower in energy than the triplet, which results from an antiferromagnetic interaction between the two electrons. A negative value of J (the singlet is higher in energy than the triplet) is typical of a ferromagnetic interaction.

The Pauli exclusion principle states that two electrons with parallel spins cannot occupy the same orbital. This keeps electrons with parallel spins apart which reduces the Coulombic repulsion between them. Systems with parallel and antiparallel spins thus have different energies. This is the origin of the exchange interaction.

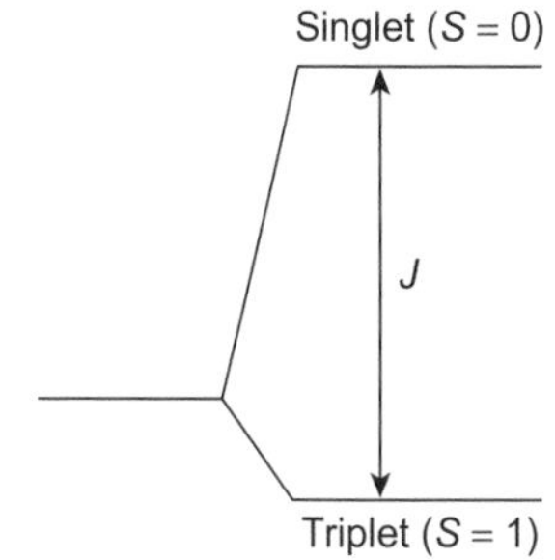

Fig. 7.1 Exchange interaction splits the energy of the system into triplet and singlet states. The diagram corresponds to a negative J (ferromagnetic interaction).

Exchange coupling requires the overlap of the orbitals containing the two electrons. It is therefore a short-range interaction that decays exponentially with the distance between the electrons and becomes negligible at distances above about 15 Å. At very short distances between the electrons, the exchange interaction can be very strong, the singlet–triplet separation very large and the energy used in EPR experiments insufficient to excite singlet–triplet transitions. Hence, in such systems only the triplet state would contribute to the EPR spectra, which are dominated by the anisotropic dipole-dipole interaction, as considered in section 7.3 and the following sections.

At longer distances between the unpaired electrons, the exchange coupling J is small and comparable to the hyperfine constant a (expressed in energy units) and the anisotropic interaction is weak. The EPR spectra of such systems are affected by singlet–triplet mixing. This is often observed in organic diradicals.

EPR spectra of diradicals

Rigid organic diradicals with a small number of bonds between the unpaired electrons are characterized by strong exchange as the exchange interaction requires orbital overlap and hence operates through bonds rather than through space (Fig. 7.2). The term *strong exchange* is used when the hyperfine constant a is significantly smaller than the exchange constant J. Such diradicals show EPR spectra split by the hyperfine interaction with nuclei from both radical units; however the distance between the hyperfine lines is $a/2$ rather than a. For instance, dinitroxides with strong exchange coupling show five-line spectra with 1:2:3:2:1 intensity ratio (i.e. split by the interaction of the unpaired electron with both nitrogen nuclei ($I = 1$), see Fig. 4.6). The separation between these lines is half of that observed for similar nitroxide monoradicals (Fig. 7.3, bottom spectrum).

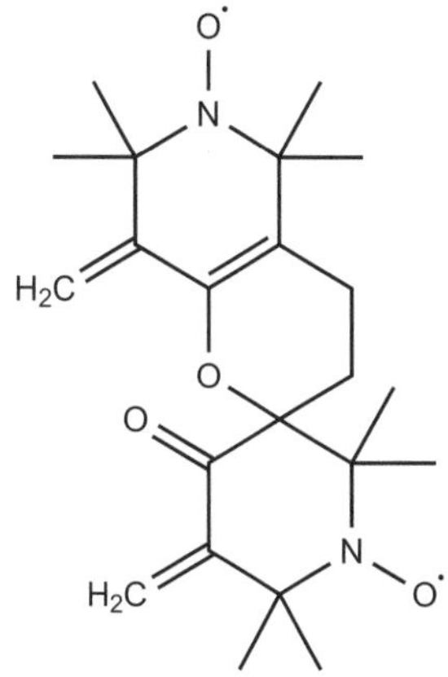

Fig. 7.2 A rigid diradical with $J >> a$.

EPR spectra of rigid diradicals with $J \approx a$ show a large number of lines; their position and intensity depend on the exact J/a ratio (Fig. 7.3). Analysis of such spectra requires a quantum mechanical treatment and computer simulations. For very weak coupling (e.g. $J << a$), the EPR spectra of diradicals are a sum of the spectra of the two monoradicals with each line exhibiting an additional splitting J, unless J is smaller than the linewidth (Fig. 7.3, top spectrum).

Many diradicals, however, are not rigid and can interconvert between alternative conformations. The orbitals of unpaired electrons might overlap in some of these conformations (thus resulting in strong exchange coupling), but are spatially separated in

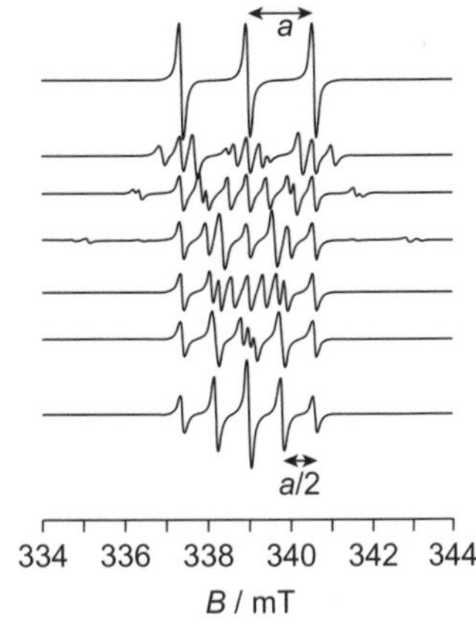

Fig. 7.3 EPR spectra of rigid diradicals, top to bottom: $J << a$, $J = a/2$, $J = a$, $J = 2a$, $J = 4a$, $J = 10a$, $J >> a$.

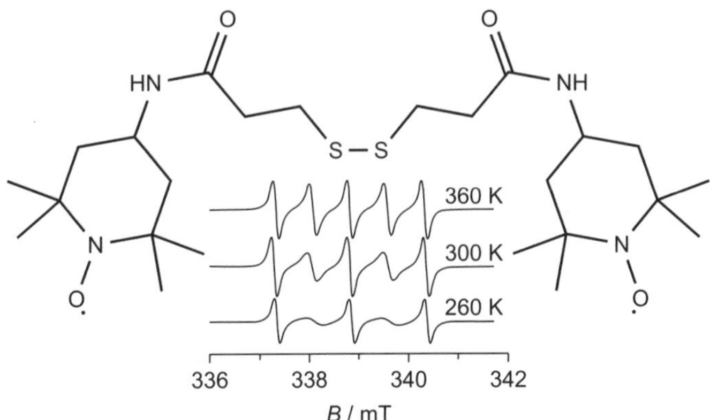

Fig. 7.4 EPR spectra of a flexible diradical at different temperatures are affected by the rate of interconversion between different conformations. The spectra in this case show alternating linewidths described in section 8.3.

others (resulting in no exchange interaction). The strength of the exchange coupling in each effective conformation, and the rates of interconversion between these different conformations all strongly affect the lineshape and linewidth (Fig. 7.4). This makes the analysis of EPR spectra complicated; however, such diradicals can be very sensitive molecular probes.

Multielectron systems: exchange broadening and exchange narrowing

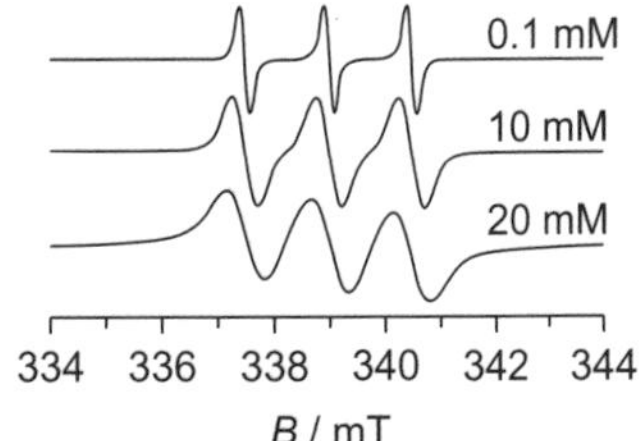

Fig. 7.5 Illustration of exchange broadening: EPR spectra of TEMPO solutions in toluene at different concentrations (see structure in Fig. 4.22).

In systems with a large number of unpaired electrons (e.g. in concentrated solutions or in neat liquids or solids), exchange interactions often lead to the disappearance of hyperfine splitting and/or changes in the lineshape. In fluid solutions, this is explained by the effect of intermolecular collisions on electron spin relaxation, as collisions of paramagnetic molecules lead to faster relaxation. This is because when two radicals collide, the orbitals with the unpaired electrons overlap momentarily and the two electrons become indistinguishable. When the radicals separate post-collision, the spin states are randomly redistributed, which is equivalent to relaxation. The increased relaxation rate results in broader EPR peaks (see section 2.5) as illustrated in Fig. 7.5. Experimental samples of compounds with narrow EPR lines should therefore be sufficiently dilute (usually below 1 mM) to avoid this *exchange broadening* (see sections 3.3 and 3.4 for advice on sample preparation).

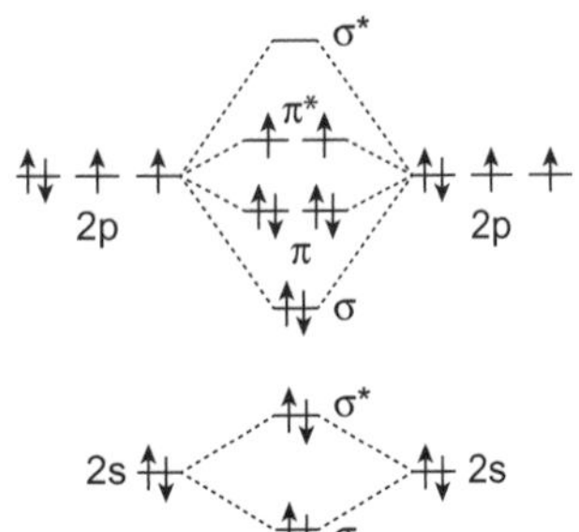

Fig. 7.6 Molecular orbital diagram for the O_2 diradical.

The sensitivity of the EPR spectral linewidth to the concentration of paramagnetic species has important applications. For instance, molecular oxygen is paramagnetic (Fig. 7.6) and hence will participate in this collisional exchange interaction. The natural concentration of dissolved oxygen in solutions is often sufficiently high to significantly broaden the EPR spectra. Therefore, well-resolved spectra of radicals with narrow EPR lines can only be obtained using degassed samples. In biomedical applications, the linewidth of EPR spectra is used to measure the concentration of oxygen in a variety of samples including tissues, cells, whole organs, and even in tumours in live mice. Any dissolved, or solid stable free radical with sharp EPR signals can be used to report on oxygen concentration (oximetry). Nitroxides are often used as soluble probes. Some charcoals (e.g. fusinite) have delocalized unpaired electrons and hence have EPR signals that are suitable for use in oximetry applications. The lithium salt of the radical anion of phthalocyanine (abbreviated as LiPc), a tetrapyrrole macrocycle structurally related to porphyrins, is another popular probe (Fig. 7.7). LiPc and charcoals give sharp EPR signals (which makes them particularly suitable for oximetry) due to exchange narrowing, explained later in this section. The experimentally measured EPR linewidth of these probes is linearly proportional to the oxygen concentration in the sample. EPR oximetry is also used in imaging mode (*EPRI* or *EPR imaging*), and provides a sensitive method for mapping oxygen concentration in biological systems.

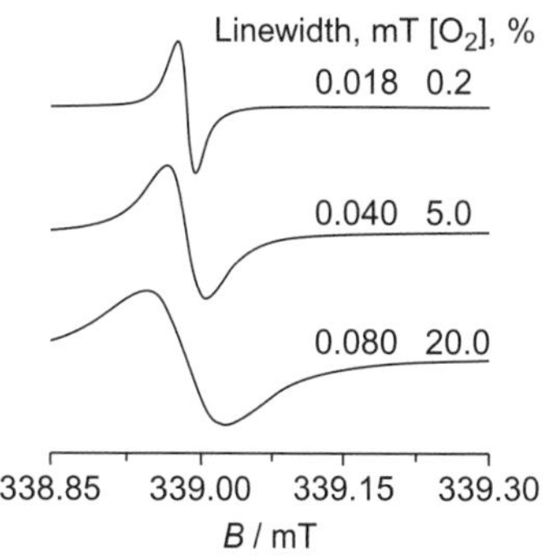

Fig. 7.7 EPR spectra of solid lithium phthalocyanine (LiPc) suspended in water with different concentrations of dissolved oxygen. Note the very small linewidth.

After M. Afeworki, N. R. Miller, N. Devasahayam, J. Cook, J. B. Mitchell, S. Subramanian, and M. C. Krishna, *Free Radical Biol. Med.* 1998, **25**, 72.

Exchange broadening is often used to study the solvent accessibility of the stable radical probes. Exchange interaction depends on the orbital overlap, hence exchange broadening can only occur if the radicals physically collide. For instance, in membrane proteins reconstituted in lipid membranes, the radicals attached to a cytoplasmic (i.e. hydrophilic) part are exposed to aqueous solutions and their EPR spectra will be broadened by collisions with water-soluble paramagnetic species (often called *broadening agents* or *paramagnetic quenchers*, since at high concentration they can broaden the EPR signal beyond detection, i.e. quench it). On the other hand, radicals attached to a hydrophobic part of the protein will not collide with water-soluble quenchers and hence their EPR spectra will not be broadened. Transition metal complexes such as potassium tris(oxalatochromate) (abbreviated as CrOx for chromium oxalate) or nickel-EDDA complex (Ni-EDDA) are often used as water-soluble paramagnetic quenchers, whereas dissolved oxygen (which is much more soluble in non-polar environments than in water) is often used as a hydrophobic quencher (to probe accessibility in hydrophobic environments).

CrOx is a paramagnetic Cr(III) (d^3, $S = 3/2$) complex $K_3Cr(C_2O_4)_3{\cdot}3H_2O$ where $C_2O_4^{2-}$ is an oxalate anion.

Ni-EDDA is a paramagnetic Ni(II) (d^8, $S = 1$) complex with ethylenediamine-*N*,*N′*-diacetic acid.

At very high concentrations of radicals, EPR spectra lose hyperfine structure and collapse into a single broad line which then narrows at very high concentrations (*exchange narrowing*, Fig. 7.8). The hyperfine collapse occurs when the exchange frequency significantly exceeds the frequency difference between the hyperfine lines. In simple terms, hyperfine collapse occurs when the unpaired electron spin relaxes so fast that it does not have time to couple to the nuclear spin.

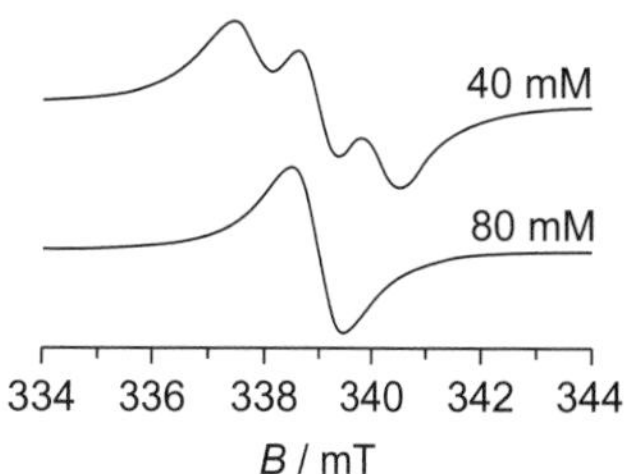

Fig. 7.8 EPR spectra of TEMPO solutions in toluene at high concentrations (see also Fig. 7.5).

In immobile solid radicals, the permanent overlap of singly-occupied molecular orbitals also leads to narrow, single-line EPR spectra which do not show hyperfine interaction. In the classical approach, one can consider the unpaired electron in these systems as being delocalized over a large area due to exchange. This delocalization means that the electron only 'experiences' the average magnetic field, and local hyperfine fields (created by the nearby nuclei) average out to zero. For example, Fig. 7.9 shows an EPR spectrum of DPPH (structure shown in Fig. 4.22) as a pure solid and as a dilute solution. The solution spectrum gives a poorly resolved multiplet arising from the hyperfine interaction with two non-equivalent nitrogen atoms; however the hyperfine structure is not seen in the solid spectrum, which gives just one narrow line. Another example is pure solid LiPc (Fig. 7.7) which also shows no hyperfine interactions with the pyrrole nitrogens. As this exchange narrowing effect requires overlap of orbitals with the unpaired electrons in adjacent radicals, it strongly depends on the nature of the radical and the packing in the solid state. For instance, many powders of pure transition metal complexes with bulky ligands do not pack well and therefore do not show exchange narrowing in their EPR spectra.

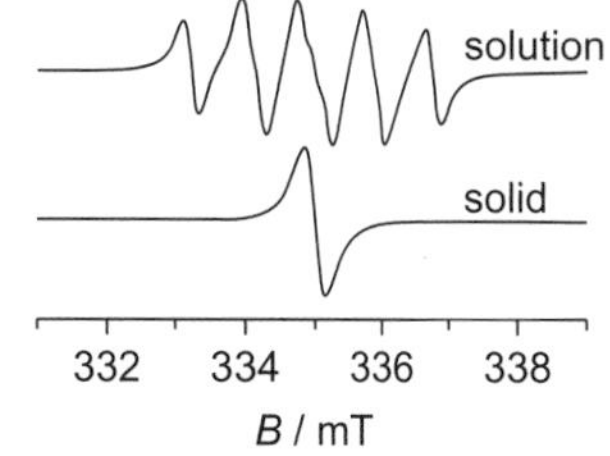

Fig. 7.9 EPR spectra of DPPH recorded in solution and as a solid.

A special case of high concentration of unpaired electrons in solids occurs for conducting materials, e.g. metals. Here the hyperfine structure is also not observed, although this is due to the rapid movement of conduction electrons through the

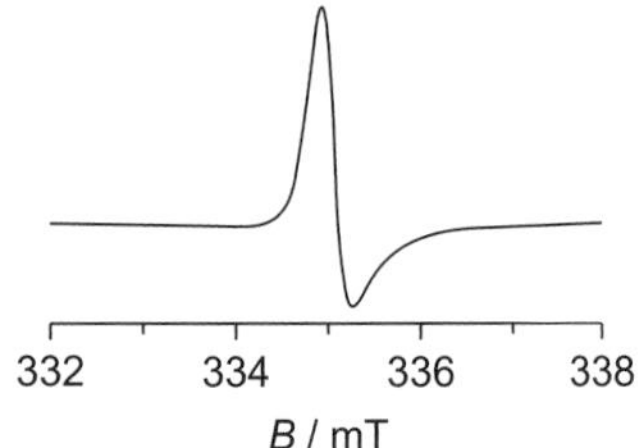

Fig. 7.10 A typical Dysonian lineshape of an EPR spectrum of a metal.

material rather than exchange interaction. The EPR spectra of conduction electrons show the characteristic asymmetric *Dysonian* lineshape (Fig. 7.10), which arises from limited penetration depth of the microwave field into the conducting sample. For heavy element metals, spin-orbit coupling leads to very fast relaxation; this broadens the conduction electron peak, sometimes so much that it becomes impossible to detect.

7.3 Dipole-dipole interaction

Theory of dipole-dipole interaction

The dipole-dipole interaction between two unpaired electrons is analogous to the dipolar contribution to the hyperfine interaction (see section 5.4) and can be considered the effect of the magnetic field created by one electron spin on the other. The EPR patterns which result from the dipolar interaction are sometimes called *fine structure*, although this term is not as commonly used as the term hyperfine structure. The dipole-dipole interaction is anisotropic and depends on the distance between the electrons and their relative orientation, similarly to the dipolar interaction between the electron and nuclear spins in the hyperfine interaction (cf. Fig. 5.13). The corresponding term in the spin Hamiltonian (eqn 7.2) is also similar to that for the dipolar contribution to the hyperfine interaction (eqn 2.21).

$$\hat{H} = \frac{\mu_0}{4\pi} g_1 g_2 \mu_B^2 \left[\frac{\hat{\mathbf{S}}_1^T \cdot \hat{\mathbf{S}}_2}{r^3} - 3 \frac{(\hat{\mathbf{S}}_1^T \cdot \boldsymbol{r})(\hat{\mathbf{S}}_2^T \cdot \boldsymbol{r})}{r^5} \right] \tag{7.2}$$

Here $\hat{\mathbf{S}}_1$ and $\hat{\mathbf{S}}_2$ are the electron spin angular momentum operators for the two electrons, and $\boldsymbol{r}$ is the interconnecting vector. The spin Hamiltonian term can be transformed to the following form:

$$\hat{H} = \hat{\mathbf{S}}_1^T \cdot \mathbf{D} \cdot \hat{\mathbf{S}}_2 \tag{7.3}$$

The splitting of energy levels into singlet and triplet states is explained by the exchange interaction, Fig. 7.1. The triplet state is then further split by the dipole-dipole interaction.

where $\mathbf{D}$ is termed the dipolar tensor. Diagonalization of this traceless tensor results in just two independent parameters D and E which derive from the principal values of the $\mathbf{D}$ tensor and characterize the dipole-dipole interaction:

$$D = \frac{3}{2} D_{zz};\ E = \frac{1}{2}(D_{xx} - D_{yy}) \tag{7.4}$$

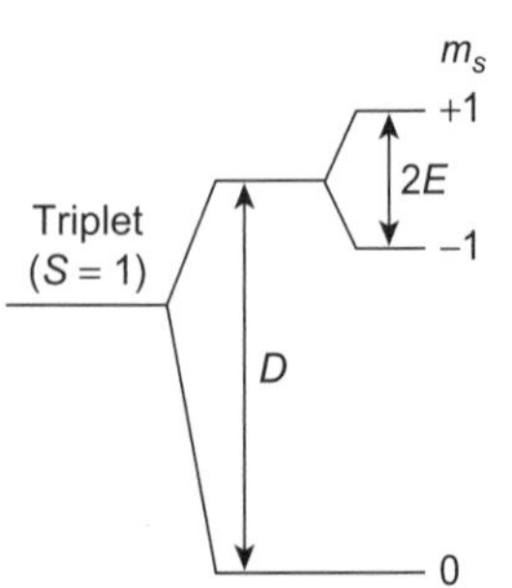

Fig. 7.11 Splitting of the triplet state by the dipole-dipole interaction between two unpaired electrons.

The dipole-dipole interaction leads to the splitting of the triplet state (i.e. $S = 1$, see Fig. 7.1) into three energy levels corresponding to the m_S values −1, 0, and +1 (Fig. 7.11). These energy levels are not degenerate in the absence of a magnetic field. The presence of more than one energy levels in the triplet state in the absence of a magnetic field is termed *zero-field splitting* (ZFS). Due to this zero-field splitting, the allowed EPR transitions between m_S levels $-1 \leftrightarrow 0$ and $0 \leftrightarrow +1$ have different energies, thus leading to two peaks in the EPR spectrum (Fig. 7.12). For systems with axial symmetry, $D_{xx} = D_{yy}$ and hence $E = 0$ (see eqn 7.4). The splitting of two energy levels in the absence of a magnetic field in this case is given only by the D parameter; hence D is sometimes called an axial ZFS parameter. The E parameter is similarly termed a rhombic ZFS parameter.

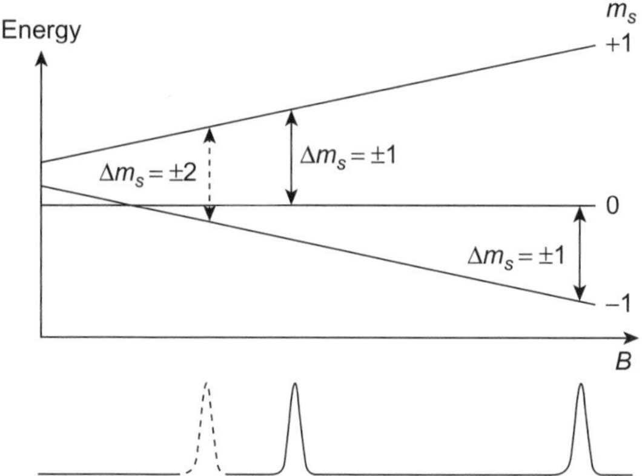

Fig. 7.12 Effect of zero-field splitting on the EPR spectrum of a triplet state. The EPR spectrum in absorption mode is shown under the energy level diagram. The $\Delta m_S = \pm 1$ transitions and the corresponding EPR peaks are shown with solid lines, the $\Delta m_S = \pm 2$ transition and the corresponding EPR absorption is shown with a dashed line.

Zero-field splitting can also be caused by other effects, e.g. spin–orbit coupling, which is observed in high-spin transition metal ions (section 7.6). The effect of these interactions is indistinguishable from that of the dipole–dipole interaction.

The resonance field for $\Delta m_S = \pm 1$ transitions predicted by eqn 2.9 is $B = \frac{h\nu}{g\mu_B}$. For $\Delta m_S = \pm 2$ transitions the resonance is observed at about half that field: $B \approx \frac{h\nu}{2g\mu_B}$. The double-photon transition is observed at the 'normal' resonance field, e.g. $B = \frac{h\nu}{g\mu_B}$.

Normally, EPR transitions are limited to the transitions with $\Delta m_S = \pm 1$. However, for triplet states at low fields a usually forbidden transition with $\Delta m_S = \pm 2$ can also be observed (Fig. 7.12). This transition, which is observed at a magnetic field approximately half of that predicted by the g value (eqn 2.9) is commonly referred to as the *half-field transition*, and is often a very diagnostic feature of systems with multiple electrons. In addition, a double-photon transition (corresponding to consecutive absorption of two $h\nu$ photons) is also sometimes observed in the EPR spectra at the expected resonance magnetic field (i.e. that predicted by eqn 2.9).

Anisotropy of dipolar interactions

The dipolar interaction between two unpaired electrons is anisotropic, and hence the EPR spectra depend on the orientation of the system with respect to the external magnetic field (similar to the anisotropic component of the hyperfine interaction, see section 5.4). The simplified two-peak (or three-peak if one takes into account the half-field transition) spectra like the one in Fig. 7.12 are only observed for samples where all molecules are in the same orientation (e.g. single crystals). In powders and glasses (e.g. frozen solutions), the orientation of molecules is random, and the resulting EPR spectra are a sum of contributions from each orientation (see section 5.3). This summation leads to very characteristic shapes for axial ($E = 0$, Fig. 7.13a) and rhombic ($E \neq 0$, Fig. 7.13b) systems. Fortunately, analysis of these spectra is simple as the values of D and E can be extracted directly from the position of the lines (in magnetic field units), as shown in Fig. 7.13. It should be noted that the half-field transition (which is nearly isotropic) is also often visible in the spectra.

In systems with well-separated unpaired electrons and small **g** anisotropy (e.g. diradicals), the dipole–dipole interaction can usually be approximated by assuming that the magnetic moments of the two unpaired electrons are parallel. This simplifies eqn 7.2 (the second term in brackets disappears), and the resulting spectra look like the one presented in Fig. 7.13(a). The integrated lineshape (shown with the dotted line in Fig. 7.13a) is known as a Pake pattern (see section 5.4).

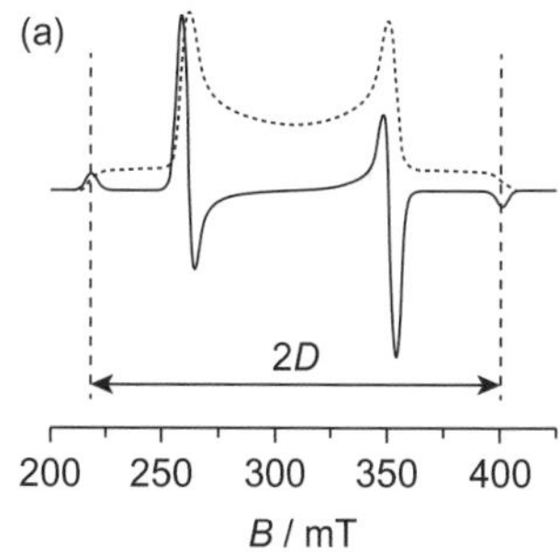

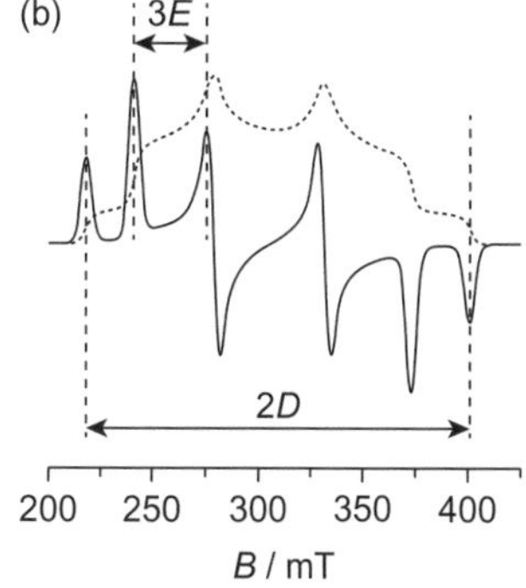

Fig. 7.13 EPR spectra of a triplet state with (a) axial and (b) rhombic symmetry in a frozen solution. Absorption profiles (dotted line) and their first integral (solid line).

7.4 Organic triplets

Non-Kekulé molecules are fully conjugated structures for which a classical Kekulé structure cannot be assigned. These molecules possess an even number of formal radical centres (e.g. 2, 4, etc.). Examples of non-Kekulé molecules are trimethylene methane or *m*-xylylene in Fig. 7.14

The dipolar interaction in triplets is quite strong and does not average out to zero even for rapidly tumbling molecules in solution. Hence the EPR spectra of solutions of organic triplets are extremely broad and are not detectable at room temperature. Therefore studies of organic triplets have to be carried out in frozen solutions or with crystals.

Organic compounds with a triplet ground state or thermally accessible triplet state are relatively uncommon, as the singlet ground state is typically much lower in energy than the triplet state in most closed shell structures. Most triplets (e.g. carbenes, nitrenes, or non-Kekulé hydrocarbons) are short lived and have to be generated in situ by photolysis or electrolysis. Examples of such structures are shown in Fig. 7.14.

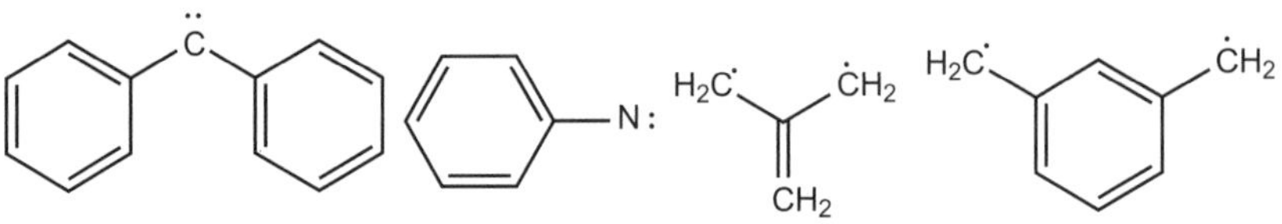

diphenyl carbene phenyl nitrene trimethylene methane *m*-xylylene

Fig. 7.14 Examples of organic molecules with triplet ground states or thermally accessible triplet states.

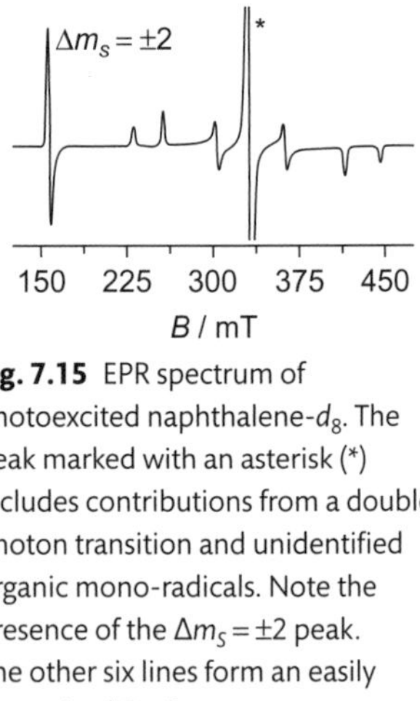

$\Delta m_S = \pm 2$ *

150 225 300 375 450

B / mT

Fig. 7.15 EPR spectrum of photoexcited naphthalene-d_8. The peak marked with an asterisk (*) includes contributions from a double photon transition and unidentified organic mono-radicals. Note the presence of the $\Delta m_S = \pm 2$ peak. The other six lines form an easily recognized $S = 1$ spectrum pattern, as illustrated in Fig. 7.13(b).

After E. Wasserman, L. C. Snyder, and W. A. Yager, *J. Chem. Phys.* 1964, **41**, *1763*.

A more common type of organic triplet is generated by photoexcitation. Absorption of UV or visible photon by a molecule in a singlet ground state leads to the formation of an excited singlet state that rapidly decays to the lowest lying excited singlet through internal conversion. Some of these short-lived singlets then undergo intersystem crossing to form a triplet state. This is commonly observed for conjugated systems (e.g. aromatic compounds), or compounds with lone pairs (e.g. ketones). The triplet states are relatively long lived. Continuous UV irradiation is often used to create a sufficiently high steady-state concentration of the triplet state that it can be studied by EPR in-situ. An example EPR spectrum of a photoexcited triplet is given in Fig. 7.15. Apart from applications to simple organic triplets, EPR studies of intermediate states in photosynthesis (which include photoexcited triplet states) have attracted much attention.

7.5 Dipolar interaction in diradicals and similar multielectron systems

The dipole–dipole interaction in systems with well-separated unpaired electrons (e.g. diradicals) is not very strong and averages out to zero for rapidly tumbling molecules in solutions. Therefore, solution EPR spectra of these systems are dominated by the exchange interaction as described in section 7.2. In frozen solutions, however, the EPR spectra can become very complex as they are affected not only by exchange interaction but also by anisotropic hyperfine interaction and anisotropic dipole–dipole interaction. For instance, the EPR spectra of frozen solutions of rigid diradicals with a short distance between the unpaired electrons have many peaks that are broad and poorly resolved and hence difficult to analyse.

Fortunately, EPR spectra of frozen solutions of diradicals are much simpler if the distance between the radicals is over about 1–1.5 nm. The exchange interaction becomes negligible at these distances for both rigid and flexible diradicals (there are no conformational changes in immobile frozen solutions, hence flexible and rigid diradicals behave in a similar way). The EPR spectra are thus dominated by the dipolar interaction. The dipolar interaction at these distances is also fairly weak, and rather than introduce additional splitting as in organic triplets (Fig. 7.13a), it only leads to some line broadening of the EPR spectra (illustrated for bis-nitroxides in Fig. 7.16). The strength of the dipole-dipole interaction, and hence the extent of the line broadening, strongly depend on the distance between the radical centres of the diradicals (eqn 7.2).

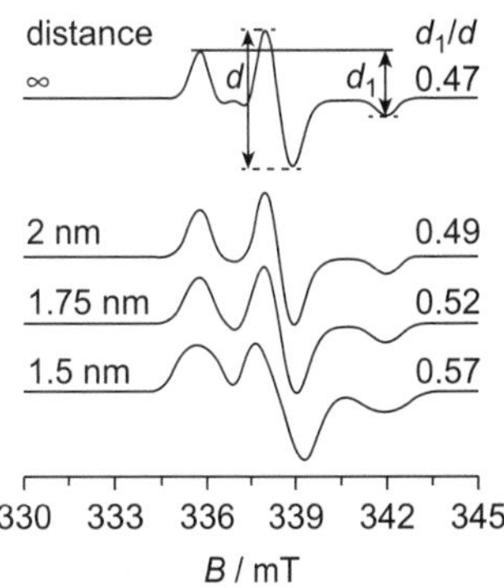

Fig. 7.16 EPR spectra of frozen solutions of mono and bisnitroxides with different distances between the radical centres. The distances correlate with an empirical ratio of peak heights d_1/d.

Although all spectra in Fig. 7.16 may look quite similar, the subtle differences in the lineshape can be reproduced very accurately by computer simulations, which makes it possible to directly estimate the distance between the two radicals in the range of about 1-2.5 nm (and up to 6-10 nm using pulsed EPR techniques also based on the dipole-dipole interaction, see Chapter 9). Alternatively, the distance between the two radicals correlates well with an empirical parameter d_1/d defined as a ratio of peak heights (Fig. 7.16). This sensitivity of EPR to the distance between radical centres in a diradical (through dipole-dipole interaction) is very important in biophysics. For instance, a protein can be specifically labelled with two stable radicals, and the distance between them measured by advanced EPR methods providing structural information and making it possible to monitor conformational changes or binding events.

In a similar way, EPR spectra of frozen solutions of monoradicals at high concentrations also show dipolar broadening which is concentration-dependent (as the concentration determines the average distance between the radical centres). Analysis of EPR spectra in these systems thus makes it possible to assess the average distance between the radicals, which is important in heterogeneous and nanostructured systems.

7.6 Transition metal ions with $S > ½$

In transition metal ions with $S > ½$, the zero-field splitting is dominated by spin-orbit coupling rather than dipole-dipole interaction. These two contributions, however, cannot be distinguished by EPR as they are characterized by identical spin Hamiltonian terms (eqn 7.2), and hence their combined effect is identical to that of just the dipole-dipole interaction described in section 7.3. Zero-field splitting in these systems can be much stronger than any other magnetic interactions, including the electron Zeeman interaction. This often shifts many EPR transitions outside of the commonly accessible range of magnetic fields, at least at X-band. High frequency/high field EPR studies are therefore often used to obtain information about electronic structure of the complexes of these ions. Unfortunately, the relaxation rates are often very fast (and hence the lines are very broad) making it difficult to observe EPR spectra at temperatures above a few kelvin. With few exceptions, the EPR spectra of fluid solutions of such ions cannot be detected at all.

Unlike most other systems discussed in this book, there are no 'typical patterns' in EPR spectra of transition metal ions with $S > ½$, partly because zero-field splitting parameters are very sensitive to molecular structure and even relatively small variations in these parameters can dramatically change the appearance of the spectra. EPR spectroscopy of $S > ½$ ions is thus a difficult field. Here, we will provide a brief introduction to the EPR spectra of transition metal ions with integer (e.g. $S = 1, 2$, etc.) and non-integer (e.g. $S = ½, 3/2$, etc.) spins, which result in quite different spectral profiles.

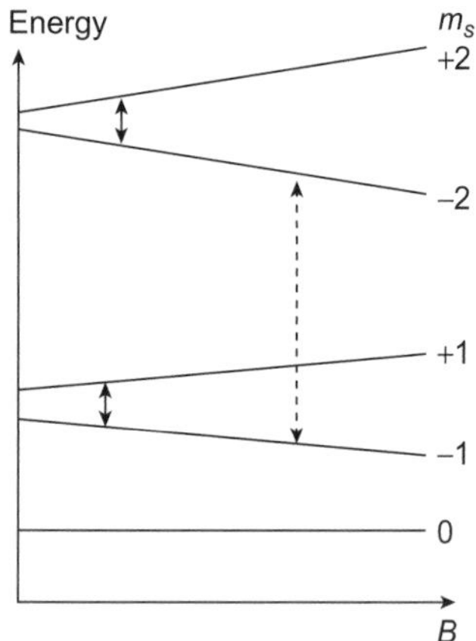

Fig. 7.17 Energy level diagram for an $S = 2$ system. Transitions within non-Kramers doublet, which can be observed at accessible microwave frequencies, are shown with solid arrows. Transitions between the non-Kramers doublets (e.g. shown with a dashed arrow) require extremely high magnetic fields and hence are usually not observed.

Ions with integer values of *S*

Examples of transition metal ions with integer values of *S* include high spin Ni(II) (d^8, $S = 1$), Fe(II) (d^6, $S = 2$), Cr(II) (d^4, $S = 2$), and V(III) (d^2, $S = 1$). A typical energy level diagram for integer *S* ions is shown in Fig. 7.17. The number of energy levels is $2S + 1$, and the energy levels with opposite values of m_S (e.g. $m_S = -2$ and $m_S = +2$) are grouped together. Notice that these pairs of levels (often termed non-Kramers doublets) are usually not degenerate in the absence of a magnetic field. Transitions *within* non-Kramers doublets (e.g. between energy levels with opposite m_S values, such as from $m_S = -2$ to $m_S = +2$, see solid arrows in Fig. 7.17) are possible and these transitions are not forbidden for very strong zero-field interactions. However, transitions *between* the doublets (such as from $m_S = -1$ to $m_S = -2$, see dashed arrow) are usually not possible as the energy of these transitions is too high to be observed using microwave frequencies accessible in EPR.

In many cases, the EPR spectra of transition metal ions with integer *S* are extremely broad and very weak and so often cannot be observed at all. The EPR spectra of these systems are often recorded in parallel mode (i.e. transitions are induced by the magnetic component of the microwave irradiation parallel to the external magnetic field), using specially designed resonators. Even when these spectra can be observed, they usually cannot be interpreted without computer simulations. Fig. 7.18 shows an EPR spectrum of a Ni(II) complex which has axial symmetry ($E = 0$). This is a rare example of an easy-to-interpret spectrum of a transition metal with integer *S*, but the lines are very broad even in this case, and not all transitions are visible.

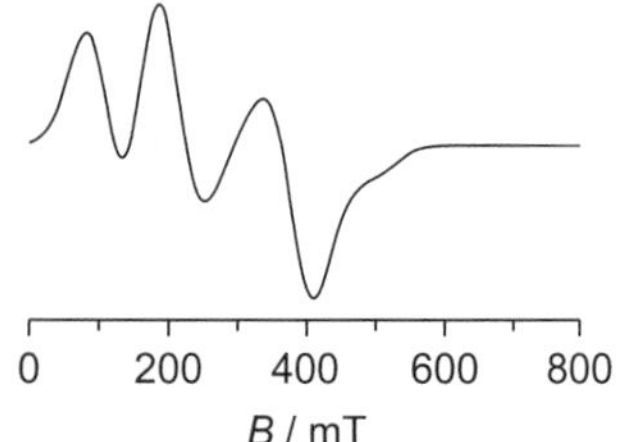

Fig. 7.18 An EPR spectrum of $Ni(CH_3CN)_6(GaCl_4)_2$. The low field line is the $\Delta m_S = \pm 2$ peak, and the other two peaks are the central two lines of the Pake pattern (Fig. 7.13a). The two smaller components of the Pake pattern are not visible. Note the broad linewidth.

After J. Reedijk and B. Nieuwenhuijse, *Recl. Trav. Chim. Pays-Bas.* 1972, **91**, 533.

EPR spectra of ions with half-integer values of *S*

Typical examples of transition metal ions with half-integer electron spin are high spin Co(II) (d^7, $S = 3/2$), Fe(III) and isoelectronic Mn(II) (d^5, $S = 5/2$), Cr(III) (d^3, $S = 3/2$). The energy levels for transition metal ions with an odd number of unpaired electrons (e.g. with half-integer *S*) group into pairs which are degenerate in the absence of an applied magnetic field. These pairs of energy levels with the opposite values of m_S are called Kramers doublets. Usually (if the zero-field splitting is much stronger than the Zeeman interaction) only transitions *between* the levels within the doublets (e.g. from $m_S = -½$ to $m_S = +½$) are observed in the EPR spectra. These transitions within the Kramers doublet resemble systems with just one unpaired electron (e.g. $S = ½$), and so Kramers doublets are sometimes termed systems with an effective spin $S' = ½$. The effective *g* values for these transitions are anisotropic and depend on the ratio of the zero-field splitting parameters *E/D*; they cover a very wide range and can be calculated using computer simulations.

For systems with axial symmetry (i.e. $E = 0$), only transitions within the first Kramers doublet (from $m_S = -½$ to $m_S = +½$) are observed. However, with increased rhombicity (i.e. an increased *E/D* ratio) transitions in other Kramers doublets gain intensity. The EPR spectra thus show a combination of transitions in each Kramers doublet (Fig. 7.19). Some features are very characteristic. For instance, the three effective *g* values for the $m_S = -3/2$ to $m_S = +3/2$ transition in Fe(III) in a rhombic environment ($E/D > 0.3$) coincide and give an intense peak with an effective *g* value of 4.3. This peak appears in many biological samples. The other visible (but less intense) transition for rhombic Fe(III) is in the Kramers doublet from $m_S = -½$ to $m_S = +½$ at $g \approx 9.5$.

In some rare cases, zero-field splitting in transition metals could be small (compared to the Zeeman interaction). A notable example is Mn(II) ($S = 5/2$), another common component in biological samples. Mn(II) shows a very characteristic EPR spectrum which is dominated by $m_S = -½$ to $m_S = +½$ transitions. The **g** and hyperfine tensors are nearly isotropic and hence the solution and solid state spectra have similar shapes. Solution spectra usually show just six strong lines due to the hyperfine interaction ($I(^{55}Mn) = 5/2$); frozen solution spectra show some weaker additional lines between the six strong hyperfine lines due to forbidden transitions ($\Delta m_I = 1$). All other transitions often overlap to give a very broad background (Fig. 7.20).

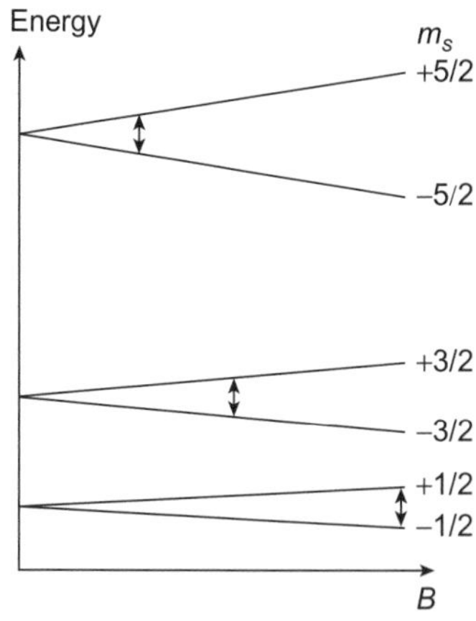

Fig. 7.19 Energy level diagram for an $S = 5/2$ system. Transitions within the Kramers doublet are shown with arrows.

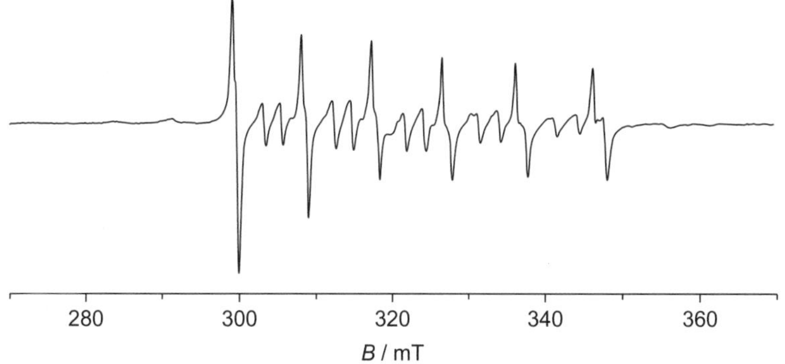

Fig. 7.20 EPR spectrum of Blu-tack shows a strong Mn(II) ($S = 5/2$) signal just like many other clays and plasticine. Very weak zero-field splitting and nearly isotropic **g** and **A** tensors result in a simple spectrum which is dominated by $m_S = -½$ to $m_S = +½$ transitions split by the hyperfine interaction.

7.7 **Summary**

- There are two types of interactions between unpaired electrons: exchange (isotropic) and dipole-dipole (anisotropic).
- In systems with a relatively large distance between unpaired electrons (e.g. diradicals or concentrated solutions of monoradicals), the exchange interaction dominates in fluid solutions, but the dipole-dipole interaction dominates in frozen solutions.
- Both interactions broaden EPR spectra. Analysis of this line broadening can yield valuable information about the concentration and accessibility in fluid solutions, and distance between the radicals in frozen solutions.
- In systems with a short distance between unpaired electrons, anisotropic zero-field splitting (caused by either dipole-dipole interaction or spin-orbit coupling) is often much stronger than any other interaction (including Zeeman interaction). Lines are broad and spectra extend over a large range of magnetic fields. The spectral shape is strongly affected by small variations in molecular structure, there are often no recognizable patterns.
- Very low temperatures are required to record the EPR spectra, and in some cases (e.g. transition metal ions with integer spin S) the lines can be too broad to be detected at all.

7.8 Exercises

7.1) Open a triplet spectrum at www.EPRsimulator.org and measure D (and E, if appropriate) values. Express your answer in energy units.

7.2) Sketch the EPR spectrum of a fluid solution of the diradical in Fig. 7.2 with both nitrogen atoms replaced by the ^{15}N isotope. Pay particular attention to the number of lines, their relative intensity, and the distance between the lines.

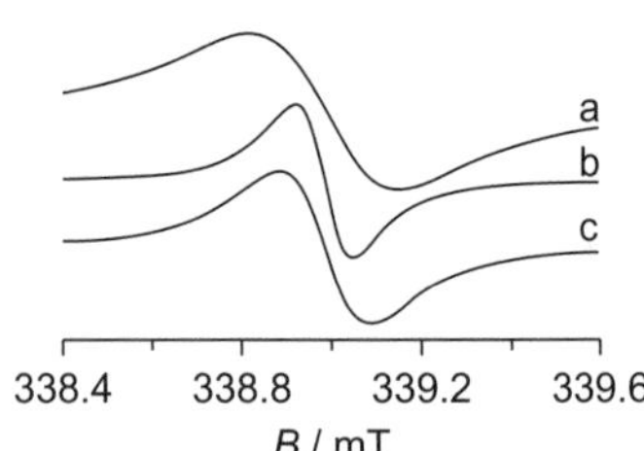

Fig. 7.21 An expansion of the EPR spectra of a TEMPO solution in acrylic acid which was (a) oxygen-saturated, (b) completely deoxygenated, and (c) contains an unknown amount of oxygen.

7.3) Fig. 7.21 shows an expansion of the EPR spectra of a dilute solution of TEMPO in acrylic acid (only one peak is shown) which was oxygen-saturated, deoxygenated, or contains an unknown concentration of oxygen. Estimate the concentration of oxygen (in g/ml) in the unknown sample, assuming that the solubility of oxygen in acrylic acid is 1.5 g/ml.

7.4) Sketch the EPR spectrum expected from the naphthalene-d_8 triplet state (see Fig. 7.15) at Q-band.

7.5) A colleague asks your advice on recording an EPR spectrum of a V(III) complex. Discuss the feasibility of this experiment, what information can be obtained from the experiment, and how to determine the best conditions for recording the spectrum.

8 Linewidth of EPR spectra

8.1 Introduction

Linewidth is a very important parameter in EPR spectroscopy. Very narrow lines can be easily saturated or distorted if incorrect spectrometer parameters (e.g. power, time constant, or modulation amplitude) are chosen (see Chapter 3). On the other hand, if recorded correctly, narrow lines offer excellent resolution of adjacent peaks with high sensitivity. Broad lines can be difficult to distinguish from the background, particularly if the intensity is not very high, and in some cases very broad lines cannot be detected at all. Linewidth measurements also provide some important information about the structure and dynamics of the spin system, which is discussed in this chapter.

In many areas of analytical science, the linewidth of a peak is usually quantified as the full width at half maximum (FWHM) (Fig. 8.2). This parameter, however, is difficult to measure for first derivative lines which are recorded in CW EPR. Therefore, the linewidth in EPR spectroscopy is often given as a peak-to-peak linewidth (ΔB_{pp}, Fig. 8.2). For complex spectra (e.g. anisotropic patterns), it may not be possible to extract the linewidth directly, but information about linewidth can be obtained through spectral simulations.

Chapter 2 introduces the concepts of relaxation and shows that the linewidth in EPR spectroscopy is usually determined by the spin–spin relaxation time T_2. There are many relaxation pathways and some are quite complex. This chapter will only deal with a few important relaxation mechanisms that contribute to the linewidth.

EPR linewidths can range from extremely narrow to so broad that they cannot be detected at all. Examples of a fairly broad and a very narrow EPR spectra are shown in Fig. 7.18 (the linewidth around 70 mT) and in Fig. 8.1 (the width of the central line of only 1.5 μT), respectively.

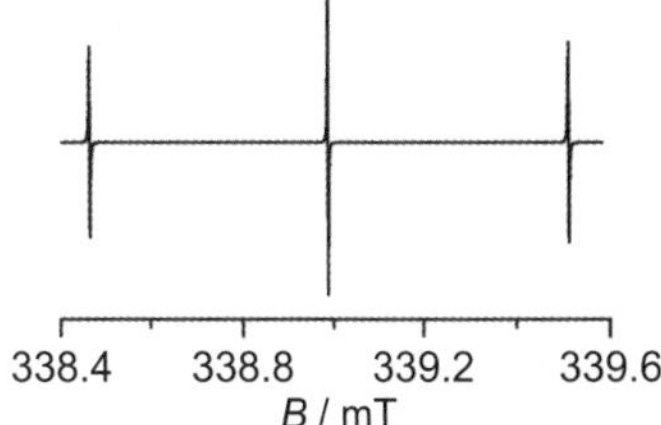

Fig. 8.1 An EPR spectrum of a ^{14}N atom ($I = 1$) trapped in C_{60}. The three hyperfine lines have different heights due to a second-order effect (not covered in this book).

After C. Knapp, K.-P. Dinse, B. Pietzak, M. Waiblinger, and A. Weidinger, *Chem. Phys. Lett.* 1997, **272**, 433.

8.2 Types of line broadening

Different types of line broadening effects are considered in sections 8.3–8.5. They can be classified as either homogeneous or inhomogeneous broadening. Homogeneous broadening arises when all spins under consideration experience identical net magnetic fields (a combination of $\boldsymbol{B}$ and $\boldsymbol{B}_{local}$ due to nearby dipoles). The transition probability as a function of the magnetic field is therefore the same for each spin in the spin packet. For instance, exchange broadening in solution (see section 7.2) provides a homogeneous contribution to the linewidth, as all molecules participating in the exchange interaction are in identical environments and hence produce identical spectra. An exclusively homogeneously broadened EPR line has a Lorentzian shape (Fig. 8.3), which is related to T_2 as follows:

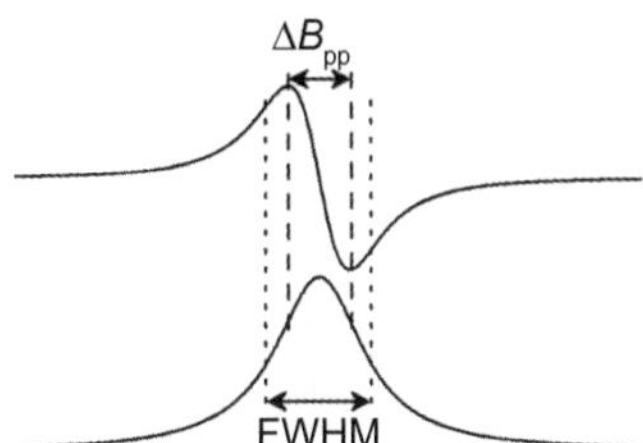

Fig. 8.2 Peak-to-peak linewidth (ΔB_{pp}, dashed lines) and FWHM (dotted lines) for a first derivative (top) and absorption (bottom) lines.

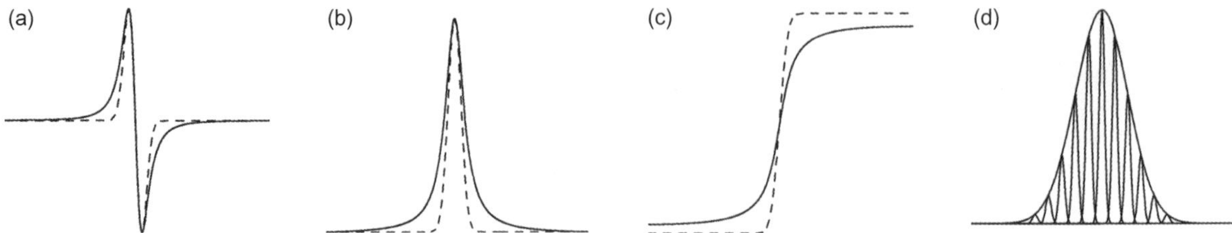

Fig. 8.3 (a) First derivative, (b) absorption, and (c) integrated absorption for Lorentzian (solid) and Gaussian (dashed) lines with the same peak-to-peak, ΔB_{pp}, linewidth. (d) The Gaussian pattern is a sum of many Lorentzian lines.

$$T_2 = 2|\gamma_e\Gamma|^{-1} = 2|\sqrt{3}\gamma_e\Delta B_{pp}|^{-1} \tag{8.1}$$

where Γ is FWHM and ΔB_{pp} is the peak-to-peak linewidth.

Inhomogeneous line broadening results from each spin packet being subjected to a slightly different net magnetic field. At each magnetic field position, only one such spin packet is exactly in resonance. Therefore the resulting EPR absorption is a superposition of individual components (which each have a Lorentzian profile), that combine to give a Gaussian envelope as shown in Fig. 8.3(d) for an absorption profile. Anisotropic EPR spectra of frozen solutions for example, frequently show inhomogeneously broadened lines as randomly distributed radicals assume different orientations with respect to the external magnetic field, and the EPR spectrum is a sum of contributions from all possible orientations (see Chapters 5 and 6).

A Voigtian lineshape does not have an analytical expression, and is often approximated by a *pseudo-Voigt* function (a linear combination of Lorentzian and Gaussian profiles).

www.EPRsimulator.org provides interactive tutorials (on the isotropic spectra page) designed to aid the familiarization with EPR lineshapes.

In many real systems, EPR lines are neither purely Lorentzian nor Gaussian but can be described by a Voigtian shape, which is a convolution of Lorentzian and Gaussian profiles.

As CW EPR spectra are recorded as first derivatives, some features of the lineshape have important implications for quantitative EPR measurements. For example, the double integral of an EPR peak is a measure of the spin concentration (see section 3.3). This is because the first integration of an EPR line (Fig. 8.3a) yields an absorption line (Fig. 8.3b), and the second integration results in the conventional integral curve (Fig. 8.3c). However, Lorentzian resonance lines have very broad wings, and a very wide field range needs to be integrated to obtain an accurate value of the double integral. It should be noted that the magnetic field axes in Fig. 8.3(a–c) are identical, and although it appears from Fig. 8.3(a) that both Lorentzian and Gaussian peaks are completely encompassed within the spectral width, Fig. 8.3(c) clearly shows that the double integral of the Lorentzian (solid) line is noticeably smaller compared to the Gaussian (dashed) line (the double integrals taken across an infinite field range of fields would be identical for both lines).

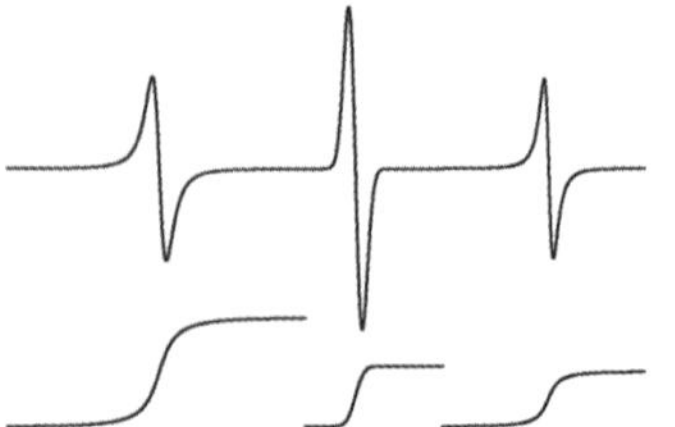

Fig. 8.4 First derivative and its double integral for Gaussian (centre) and Lorentzian (left and right) lines illustrating why the first derivative peak height should not normally be used as a measure of intensity.

Peak heights should not normally be used as a measure of concentration in EPR. Fig. 8.4 shows a fictitious EPR spectrum which has three lines in an apparent 1:2:1 intensity ratio (originating, for example, from coupling to two equivalent $I = ½$ nuclei). However, double integration shows that the intensity ratio is in fact 2:1:1. The low field line (Lorentzian) has twice the intensity, even though it is half the height of the middle line (Gaussian) with the same peak-to-peak linewidth, and twice the intensity of the high field linewidth a similar height (the high field line has smaller linewidth). For EPR lines with the same double integral intensity and the same lineshape function, the peak height is inversely proportional to the linewidth squared (e.g. a twofold increase in peak linewidth results in a fourfold decrease in peak height).

8.3 Effect of dynamic processes on the EPR linewidth

Dynamic processes which change resonance frequencies, but do not involve spin flips, will broaden the EPR lines. For example, consider a solution of a radical that exists as two tautomers characterized by different g values (and hence different resonant frequencies). If the interconversion between the two tautomers is extremely slow, the EPR spectrum will be a sum of the spectra for each tautomer, and the linewidth will not be affected. Alternatively, if the interconversion is extremely fast, then the observed spectrum will correspond to the g value which is an average for the two tautomers (the linewidth in this case is also not affected). However, at intermediate rates of interconversion between the two tautomers, the EPR lines will be broadened. As a general rule, *dynamic processes affect the EPR linewidth if the difference in resonance frequencies is comparable in magnitude to the rate of the dynamic process* (see example in the margin). 'Comparable' can be loosely defined as within 1–2 orders of magnitude. Typical differences in resonance frequencies that can be monitored by X-band EPR are in 1–1000 MHz range (about 0.05–50 mT in field units), which makes EPR linewidths potentially sensitive to dynamic processes characterized by rates of about 10^5–10^{11} s^{-1}, or correlation times τ_c around 10^{-5}–10^{-11} s.

Consider an X-band (9.5 GHz) EPR spectrum of a radical which exists as two interconverting tautomers: X ($g = 2.0015$) and Y ($g = 2.0025$). Using $h\nu = g\mu_B B$ (eqn 2.9), the resonance field is $B \approx$ 339 mT. With the same equation, the difference in frequencies between the tautomers is $\Delta\nu = [g(Y) - g(X)]\mu_B B/h =$ 4.7 MHz $\approx 10^6$ s^{-1}. Hence the linewidth of the EPR spectra will be affected by the tautomerization if the rate of interconversion between the tautomers is around 10^5–10^7 s^{-1} (e.g. half-reaction times in the μs range).

A dynamic process which affects the linewidth is said to be on the *EPR timescale*.

Rotational diffusion (molecular tumbling)

Chapter 4 describes the isotropic EPR spectra of samples recorded in fluid solutions. In these systems, the anisotropic interactions (observed in the solid state, see Chapter 5) are completely averaged out due to rapid molecular tumbling. In many cases, however, molecular tumbling is insufficiently fast to produce isotropic spectra. In order to completely average out anisotropic interactions, the rate of molecular tumbling should be much faster than the difference in resonance frequencies for different molecular orientations. For instance, in nitroxides, the difference between A_{zz} and A_{xx}/A_{yy} is about 3 mT which corresponds to about 0.8 GHz = 8×10^8 s^{-1}. In order to produce an isotropic spectrum, the rotational correlation time, τ_c, must therefore be much shorter than 10^{-9} s (in practice, below 10^{-11} s). Such short correlation times are observed for very small molecules in solutions of low viscosity, but not for large molecules or viscous solutions. Insufficiently fast rotational diffusion therefore often leads to homogeneous broadening of EPR lines, as anisotropic interactions are not completely averaged. The broadening of EPR spectra by rotational diffusion falls into two categories, *fast* and *slow motion* as described below.

The precise definition of *correlation time* (τ_c) for a dynamic process is rather complex. In simple terms, it is a reciprocal of a rate of the process. For rotational diffusion, τ_c is the mean time required to rotate a molecule by about 1 radian.

Fast motion is observed when $1/\tau_c$ (e.g. the frequency of rotation) is larger than the difference in resonance frequencies for different molecular orientations (for nitroxides $\tau_c < 10^{-9}$ s). Visually, the EPR lines are observed at the same field position but they are broadened to different extents (and hence have different peak amplitudes). For instance, small molecule nitroxides in viscous solvents often show fast motion spectra (Fig. 8.5). Another example is the vanadyl cation (see Fig. 5.15a). For VO^{2+}, the difference between $A_\perp$ and $A_\parallel$ is about 11.5 mT (see Fig. 5.15b) which corresponds to about 0.3 GHz = 3×10^8 s^{-1}. In practice, VO^{2+} in an aqueous solution tumbles with a frequency of $1/\tau_c = 2 \times 10^{10}$ s^{-1} which is fast motion but insufficient to average out the anisotropic interactions. Therefore the asymmetric shape in Fig. 5.15(a) is caused by homogeneous broadening.

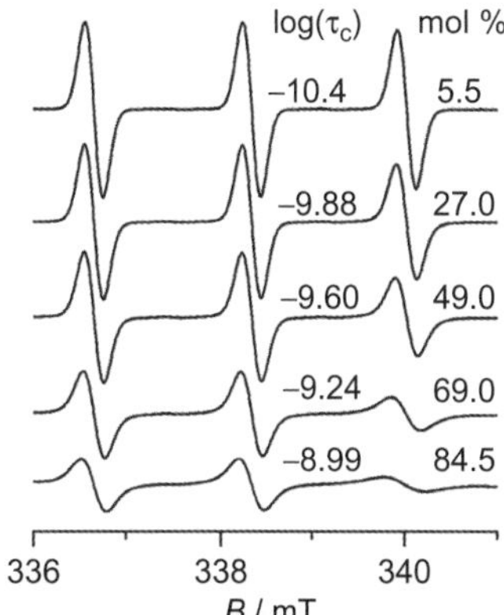

Fig. 8.5 EPR spectra of TEMPO in aqueous glycerol at 300 K (fast isotropic motion). The concentration of glycerol and the rotational correlation times τ_c are shown on the spectra. The spectra are normalized to have the same double integral intensity.

After D. Banerjee and S. V. Bhat, *J. Non-Cryst. Solids* 2009, **355**, 2433.

The precise lineshape of EPR spectra broadened by fast motion depends on whether the rotational diffusion is isotropic (e.g. rotation with the same rate in all directions) or

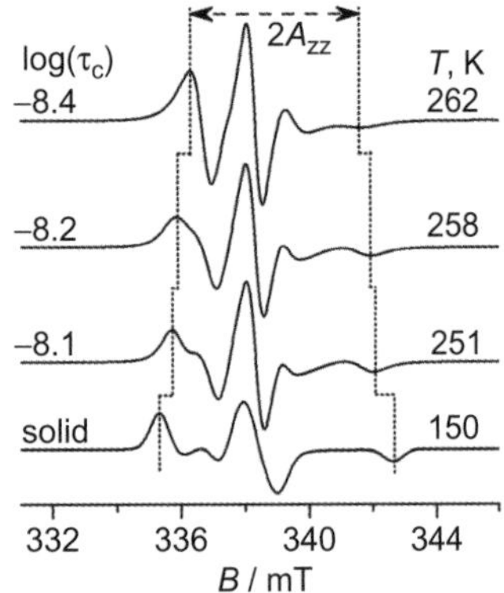

Fig. 8.6 EPR spectra of TEMPO in 69 mol % aqueous glycerol (slow isotropic motion). Temperature, rotational correlation times τ_c and effective $2A_{zz}$ values are shown on the spectra. The spectra are normalized to have the same double integral intensity.

After D. Banerjee and S. V. Bhat, *J. Non-Cryst. Solids* 2009, **355**, 2433.

anisotropic (e.g. preferrential rotation across one axis). Tumbling of flexible or nearly spherical molecules, for example TEMPO (Fig. 8.5), is well described by isotropic rotation, and the correlation time τ_c in this case can be calculated from the relative heights of the peaks. For nitroxides, eqn 8.2 is often used:

$$\tau_c = 6\times10^{-9}\frac{\text{s}}{\text{mT}}\cdot\Delta B_{pp}\cdot\left(\sqrt{\frac{h_0}{h_{-1}}}+\sqrt{\frac{h_0}{h_1}}-2\right) \qquad (8.2)$$

Here ΔB_{pp} is the peak-to-peak linewidth of the central line, h_{-1}, h_0, and h_1 are the peak-to-peak amplitudes for the high, central, and low field lines, respectively. Alternatively, rotational correlation times could be obtained from spectrum simulations.

Slow motion is observed when $1/\tau_c$ is similar to or smaller than the difference in resonance frequencies for different molecular orientations. The EPR lineshapes corresponding to slow motion are complex and depend strongly on the type of rotational diffusion (Fig. 8.6). Slow motion EPR spectra are often observed in viscous solvents, or when a radical is attached to a large molecular assembly (e.g. a colloidal system) or a macromolecule (e.g. a protein). The shape of the EPR spectrum is thus affected by the tumbling of the whole system, in addition to the local motion of the spin-bearing functional group if it is conformationally flexible. For nearly immobile samples (e.g. solids at room temperature), EPR spectra can also be significantly affected by *libration* (i.e. small amplitude, oscillatory rotation of the molecules, or groups of atoms, back and forth). A number of different models have been developed to simulate such spectra, with some using molecular dynamics to predict the type of local motion. Although such models are readily available (e.g. using EasySpin at www.easyspin.org), accurate simulation of slow motion spectra is not straightforward, particularly without precise knowledge of the types of motion.

EasySpin (www.easyspin.org) can be used for simulating EPR spectra with incomplete averaging of **A** and **g** tensors (the *garlic* function is used for isotropic fast motion, and the *chili* function for other motions).

Without simulations, it is difficult to extract dynamic information from slow motion EPR spectra. For nitroxides, the distance between the outer lines (the 'effective $2A_{zz}$ value', Fig. 8.6) is sometimes used as a qualitative measure of rotational diffusion. For isotropic motions, this parameter increases with increased τ_c, but this dependence is not linear.

The rotational correlation times τ_c of the radical as a whole (e.g. ignoring any local motion of the spin-bearing group) is related to the molecular size using the Stokes–Einstein relation:

$$\tau_c = \frac{1}{6D} = \frac{4\pi\eta r^3}{3kT} \qquad (8.3)$$

Here D is the spherical diffusion coefficient, η is the viscosity of the medium, and r is the hydrodynamic radius of the molecule modelled as a rotating sphere. EPR spectra thus provide information on molecular size. For instance, Fig. 8.5 shows that TEMPO in 49 mol % water-glycerol mixture at 300 K has $\tau_c = 10^{-9.6}$ s. The viscosity of the solvent under these conditions is $\eta = 0.063$ N s m^{-2}. Solving eqn 8.3, we get $r \approx 0.16$ nm which compares well with the TEMPO dimensions ($r \approx 0.2$ nm).

The symbol for the diffusion coefficient D should not be confused with the analogous symbol for the zero-field splitting parameter D (see section 7.3).

Chemical exchange

The treatment of chemical exchange in EPR and NMR is very similar (see Chapter 4 in the *Nuclear Magnetic Resonance* Oxford Chemistry Primer for a detailed description).

In the field of EPR, the term chemical exchange is applied to any reversible chemical or physico-chemical process where the starting material and product have different resonance frequencies. This could be a conformational transition (e.g. ring flipping of a cyclohexyl ring), a reversible chemical reaction such as tautomerization (see the example in section 8.3), a proton/ion/ligand exchange, or a hindered rotation. The line broadening in solutions of flexible diradicals for example (see Fig. 7.4) is caused by conformational transitions.

The effect of chemical exchange on EPR spectra is similar to that observed in NMR spectroscopy. At the slow rates of interconversion, the EPR spectra are a sum of those of the starting material and the product. With the increased rate of interconversion, the lines broaden, then coalesce, and finally sharpen to yield a spectrum corresponding to the averaged resonance frequencies (Fig. 8.7). The information about the rates of exchange can be obtained by simulation. When the peaks corresponding to the starting material and product just coalesce (as in Fig. 8.7), the rate of exchange k can be calculated according to:

$$k = \frac{\pi \Delta \nu}{\sqrt{2}} \tag{8.4}$$

Here $\Delta\nu$ is the difference in resonance frequency of the exchanging lines.

Exchange between chemically identical states often leads to the observation of alternating EPR linewidths. Consider the stannylated semiquinone radical in Fig. 8.8. The $SnClMe_2$ group undergoes intramolecular migration. Although the starting material and product in Fig. 8.8 are identical, the rate of migration affects the EPR spectra: this is a typical example of chemical exchange (Fig. 8.9). If the migration is slow, the unpaired electron is coupled to two non-equivalent hydrogens labelled H^A and H^B, and hence gives rise to a doublet of doublets in the EPR spectrum (four lines). If the migration is fast, the two hydrogens become equivalent and a 1:2:1 triplet is observed in the EPR spectrum.

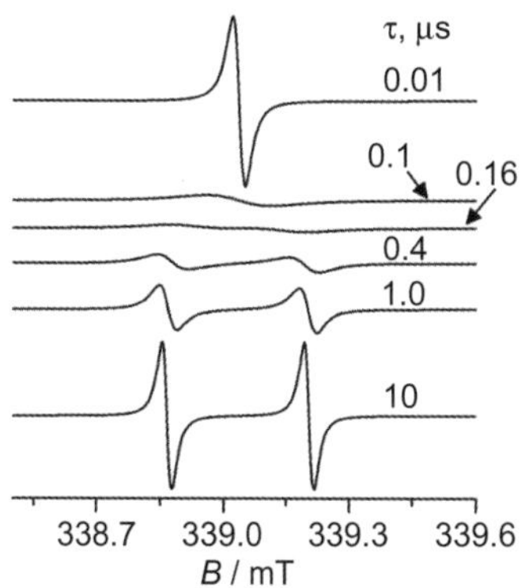

Fig. 8.7 EPR spectra of exchanging tautomers X and Y described in section 8.3. The lifetime of each tautomer is indicated. All spectra are normalized to have the same double integral intensity.

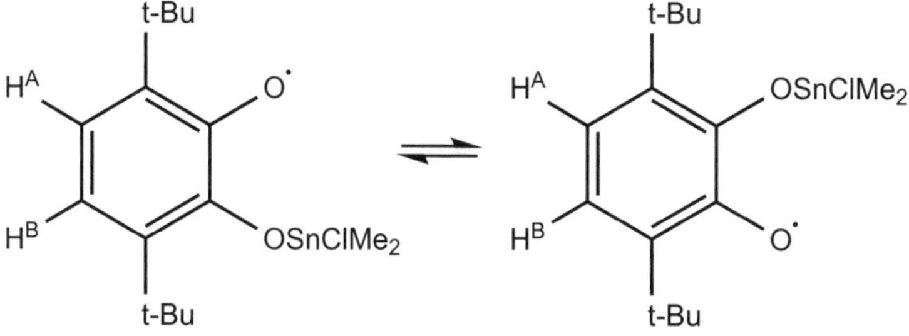

Fig. 8.8 Stannotropic migration in this phenoxyl radical leads to alternating linewidths in the EPR spectrum.

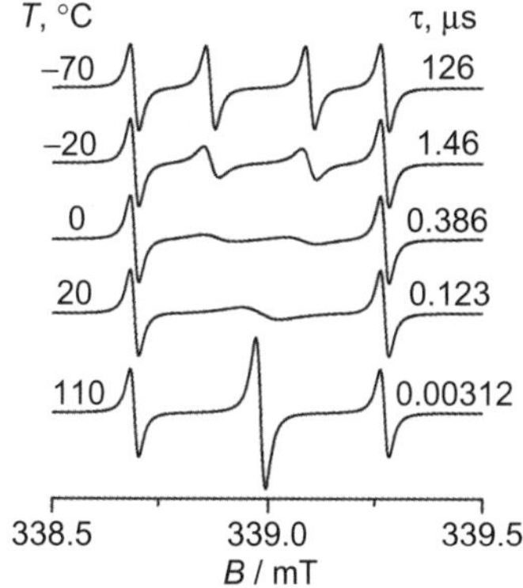

Fig. 8.9 EPR spectra of the phenoxyl radical shown in Fig. 8.8 (in THF solution). Temperatures and lifetimes are indicated.

After N. N. Bubnov, S. P. Solodovnikov, A. I. Prokof'ev, and M. I. Kabachnik, *Russ. Chem. Rev.* 1978, **47**, 549.

At intermediate rates of migration, the EPR spectrum shows two sharp lines and one or two broad lines, i.e. with alternating linewidths (Fig. 8.9). In order to understand the origin of this phenomenon, consider the alignment of the nuclear spins H^A and H^B in both structures (Fig. 8.10), either parallel (α) or antiparallel (β) to the external magnetic field. After migration, H^A and H^B swap places. One can see that the two extreme lines have the same spin alignment and hence appear at the same resonance frequencies as before migration (Fig. 8.10). As the broadening depends on the difference in resonance frequencies before and after the dynamic process, these lines are not broadened. On the other hand, the two inner lines swap position after migration, and hence are broadened if the rate of migration is comparable to the difference in

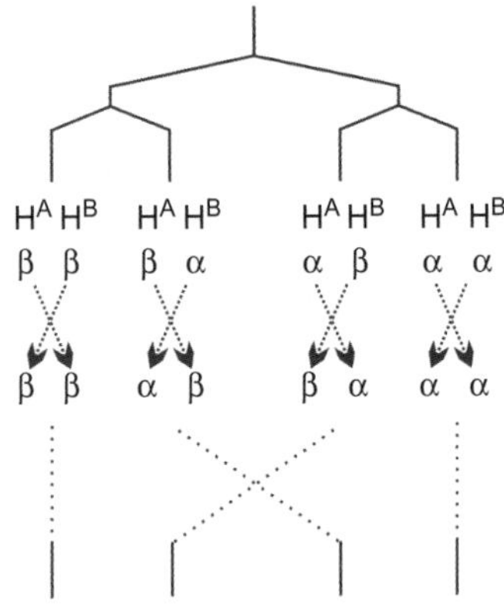

Fig. 8.10 Chemical exchange in the phenoxyl radical shown in Fig. 8.8. H^A and H^B swap places upon exchange.

the resonance frequencies before and after migration. Selective broadening of the second and fourth lines in the spectra of diradicals (see Fig. 7.4) is another example of alternating linewidths.

8.4 Lifetime of spin states

Dynamic processes which involve flipping of the electron spin reduce the lifetime of the spin states, increase the relaxation time and lead to line broadening. One example of this effect is exchange broadening, considered in section 7.2. Another similar example is electron transfer. For example, radical anions often undergo electron transfer with the parent dianions (sometimes called *self-exchange*) on the EPR timescale (Fig. 8.11). This electron transfer increases the relaxation rate and results in line broadening. The EPR spectra of radical anions therefore often depend on the concentration of the parent neutral molecule in solution. For slow exchange (e.g. when the line broadening is significantly smaller than the distance between adjacent hyperfine lines), the broadening for each line is linearly proportional to the concentration of the neutral compound. Exchange rates can thus be calculated from the linewidths, or by spectrum simulation. The broadening affects the hyperfine lines differently: the outermost lines are broadened more than the central lines. In some cases, the exchange could be very fast, and the hyperfine structure collapses, just like for exchange interaction (Fig. 8.12).

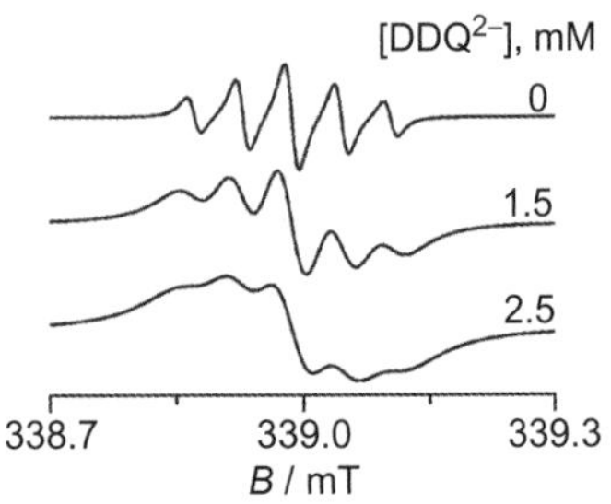

Fig. 8.12 EPR spectra of the DDQ radical anion in CH_2Cl_2 at different concentrations of DDQ dianion. The concentration of DDQ radical anion is constant for all spectra.

After G. Grampp, S. Landgraf, and K. Rasmussen, *J. Chem. Soc., Perkin Trans. 2*, 1999, 1897.

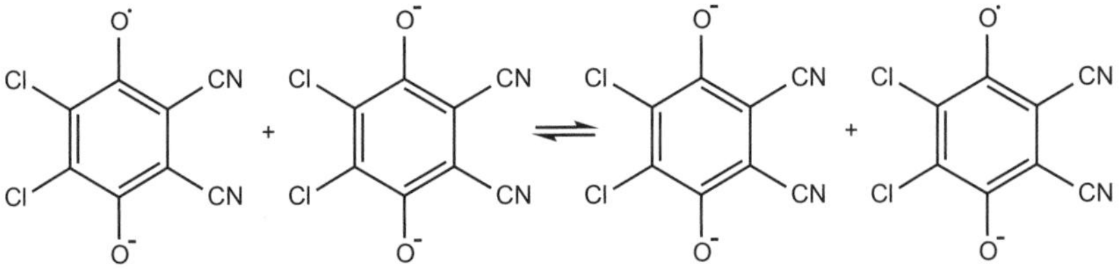

Fig. 8.11 Self-exchange of the radical anion of 2,3-dichloro-5,6-dicyano-1,4-benzoquinone (DDQ) with its dianion.

8.5 Inhomogeneous broadening

Inhomogeneous broadening is not related to dynamic processes. It is observed if different individual spins are in different environments or have different orientations with respect to the external magnetic field. EPR spectra of disordered paramagnetic species in the solid state (e.g. in frozen solutions, see section 5.3) are a sum of spectra for each orientation and hence are significantly broader than the spectra of each individual orientation. This is an example of inhomogeneous broadening. Similarly, weak dipole-dipole interactions between radicals (e.g. observed in frozen solutions of diradicals or monoradicals at high concentration, see section 7.5) make EPR spectra different for radicals in different orientations. The overall spectra which include contributions from each orientation are thus inhomogeneously broadened (see section 7.5).

Another important type of inhomogeneous broadening comes from strain effects. The term *strain* describes a small variation in magnetic parameters (e.g. components of **g**, **A**, and **D** tensors) for individual radicals due to the inhomogeneous nature of the sample, possibly caused by defects in the solid matrix or slightly different conformations in the solid phase. This is particularly important for biological samples, when rapid freezing is often used to immobilize the conformations of biomacromolecules in the solid sample. The presence of slightly different conformations in the sample leads to the distribution of magnetic parameters which usually results in the broadening of the EPR lines. While this effect may seem small, if it is not taken into account it can lead to significant errors in spectrum simulations (Fig. 8.13).

A somewhat different type of inhomogeneous broadening, not related to the sample inhomogeneity, is due to weak, unresolved hyperfine interactions. This broadening is often observed in solid samples where the close proximity of a large number of nuclei with non-zero spin (e.g. protons) yields very weak hyperfine interactions. They are usually too weak to split the EPR lines; however the combined effect of these interactions leads to line broadening. Apart from broadening, unresolved hyperfine interactions change the lineshape, making it more Gaussian (just like any other inhomogeneous broadening) (Fig. 8.14). Free radicals in solution also show this effect. For instance, TEMPO has weak hyperfine interactions with the protons in methyl and methylene groups, which affect the lineshape. Although the changes of the lineshape may seem small and insignificant, they can have surprisingly large effects on the quantitative results of any simulations, because Gaussian/Voigtian and Lorentzian lineshapes have very different double integral intensity in the wings of the lines and around the resonance frequency (see Fig. 8.3).

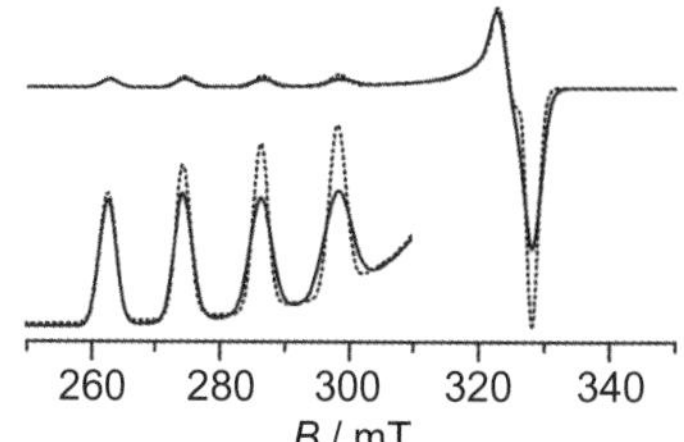

Fig. 8.13 EPR spectrum of $CuCl_2$ in water/ethanol/glycerol mixture without (dotted line) and with (solid line) correlated **g** and **A** strain.

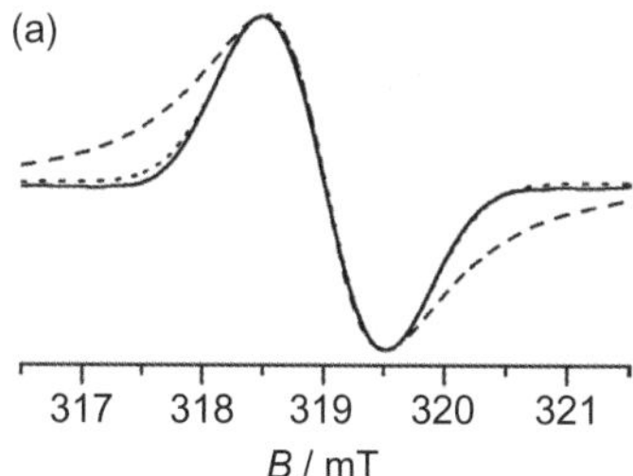

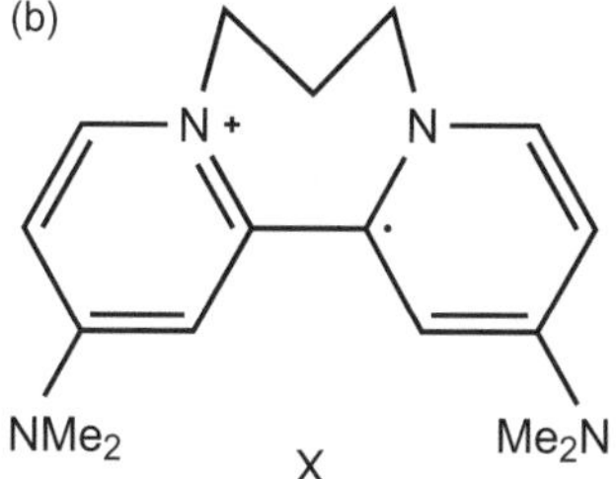

Fig. 8.14 (a) EPR spectrum of the radical cation X shows a nearly Gaussian shape due to many unresolved hyperfine interactions with four ^{14}N and twenty two ^{1}H nuclei. For comparison, Gaussian and Lorentzian shapes are shown with dotted and dashed lines, respectively. (b) Radical cation X.

8.6 Spin labels and probes

EPR spectroscopy is a technique very sensitive to the environment around the unpaired electron. Chapter 7 shows that EPR spectra report on the distances between paramagnetic centres. In this chapter, we have seen how EPR spectra can report on the rate (and anisotropy) of molecular tumbling or the dynamics of chemical exchange. Unlike other techniques (e.g. NMR), EPR spectra of even complex mixtures can be relatively simple as they only contain signals from unpaired electrons. Therefore, stable organic radicals (e.g. nitroxides) and to a lesser degree transition metal ions (e.g. vanadyl) are often used as spin probes and spin labels to investigate the properties of complex systems (Fig. 8.15). The distinction between spin probes and spin labels is somewhat blurred; spin probes usually describe radicals added to the system, whereas spin labels are usually covalently attached to a component of the system. Spin probes/labels are used to obtain information on:

- the **conformational flexibility** of the label or viscosity of the environment (from the tumbling rates);
- the **solvent accessibility** or concentration of paramagnetic species such as oxygen (from exchange broadening);
- the **distances** between two labels (from dipole–dipole interaction);
- the **polarity** of the environment (^{14}N hyperfine of a nitroxide is sensitive to polarity, Fig. 8.16) and sometimes the **chemical properties** (e.g. pH).

Spin probes/labels are particularly appropriate for studying supramolecular and colloidal systems, nanostructures, and biomacromolecules.

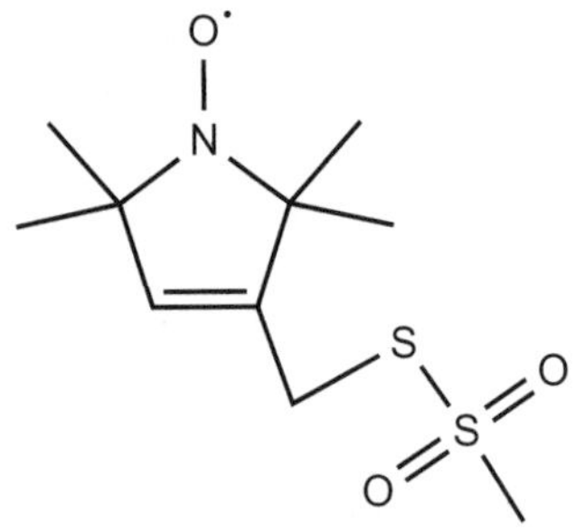

Fig. 8.15 A popular MTSSL spin label which reacts selectively with Cys aminoacids in proteins: $RSH + R'SSO_2Me \rightarrow RSSR' + MeSO_2H$.

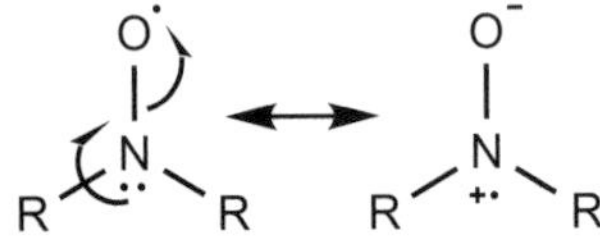

Fig. 8.16 Resonance stabilization of nitroxides. Polar solvents favour the zwitterionic form, which has a radical centre on the nitrogen atom, and hence higher $a(^{14}N)$ than the neutral form.

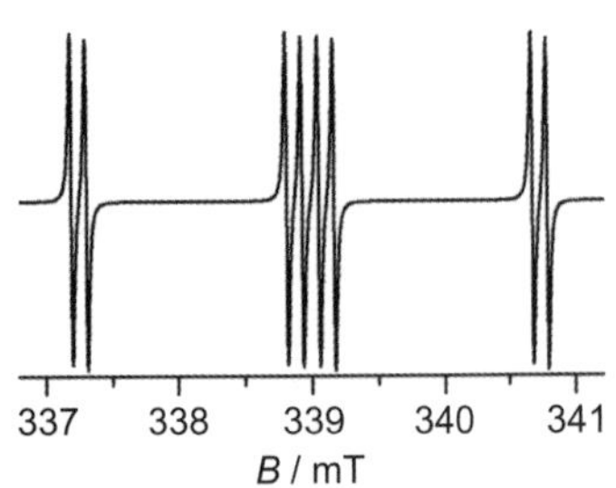

Fig. 8.17 EPR spectrum of methanol radical $HOCH_2^{\bullet}$ at 130 K.

8.7 Summary

- EPR linewidths vary significantly; this strongly affects the patterns of EPR spectra and complicates quantitative analysis.
- EPR linewidths provide very important structural and dynamic information about many paramagnetic systems.
- EPR line broadening can be homo- or inhomogeneous. Homogeneous broadening occurs when all radicals in the sample produce identical EPR spectra. Inhomogeneous broadening occurs when different radicals in the sample have different EPR spectra.
- Homo- and inhomogeneously broadened EPR spectra have Lorentzian and Gaussian profiles, respectively.
- Dynamic processes such as $A \rightleftarrows B$, where A and B have different EPR spectra, affect the EPR linewidth if the rate of the dynamic process is comparable to the difference in resonance frequency of A and B. Such dynamic processes include molecular tumbling, reversible chemical reactions, or conformational transitions.
- Dynamic processes such as $A \rightleftarrows B$, where A and B have different orientation of electron spin, lead to line broadening. Such processes include collisions between radicals (see section 7.2) and degenerate electron transfer between radicals and parent full shell molecules.
- Inhomogeneous broadening is often observed in frozen solution and can be additionally affected by strain. Another source of inhomogeneous broadening is unresolved hyperfine interactions.

8.8 Exercises

8.1) Open 'Isotropic EPR Spectra' at www.EPRSimulator.org, click on 'what can I learn on this page' and do the peak shapes exercises.

8.2) A rigid nanoparticle possessing a nitroxide group is 1 nm in diameter. The viscosity of water at 25 °C is 8.90×10^{-4} N s m^{-2}. Deduce if an aqueous solution of these nanoparticles will give fast or slow motion X-band EPR spectra at room temperature. What will be the answer for 3 nm particles?

8.3) Using the spectra shown in Fig. 8.9, calculate the rate of stannotropic migration of the phenoxyl radical in Fig. 8.8 at 20 °C if the spectra were recorded at a frequency of 9.5 GHz.

8.4) Fig. 8.17 shows an EPR spectrum of the methanol radical $HOCH_2^{\bullet}$ at 130 K. The spectrum reveals a small hyperfine interaction with the OH proton and stronger interactions with two non-equivalent protons of the CH_2 group. Upon increasing the temperature, rotation around the C–O bond leads to broadening of the EPR spectra. At about 220 K, the spectrum shows hyperfine interaction with two equivalent protons of the CH_2 group (in addition to the weak interaction with the OH proton). Deduce which EPR lines are broadened by this dynamic process, and hence sketch the EPR spectra which would be observed as the temperature is gradually increased from 130 to 220 K.

8.5) Open an EPR spectrum of a nitroxide with restricted tumbling at www.EPRSimulator.org and set log(D) to 10.

(a) Observe how the spectra change at different microwave frequency. Explain why the spectra appear almost isotropic at S-band (e.g. the three lines have almost the same intensity) but are strongly anisotropic at W-band.

(b) Set log(D) to 10 and the frequency to W-band. Observe how the EPR spectra are affected by A_{zz}. Explain why the spectra with large A_{zz} appear more anisotropic than those with small A_{zz}.

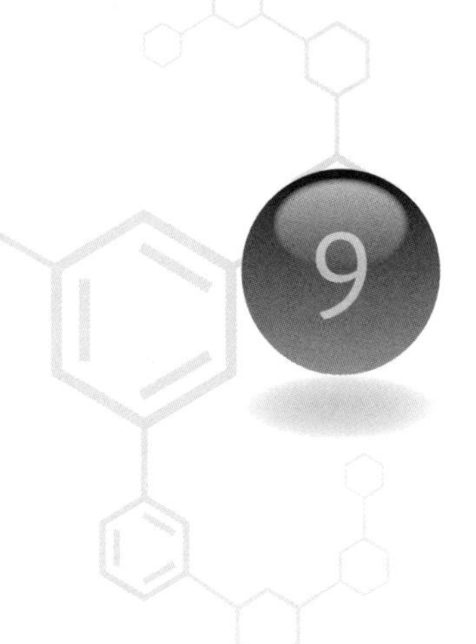

9 Advanced EPR techniques

9.1 Introduction

Chapters 2 to 8 introduce the reader to the basic theory of CW EPR, illustrating the power of the technique to study a wide range of paramagnetic systems. A great deal of the information provided by EPR spectroscopy on the structure and spin densities can be extracted through analysis of the hyperfine interactions. If the magnitudes of the couplings are small (e.g. less than the EPR linewidth), these interactions will remain unresolved in the CW EPR spectrum. Fortunately a number of advanced EPR techniques are available to recover this information and they will be introduced in this chapter. The emphasis will be placed on illustrating the additional information which can be accessed by these methods, rather than providing a detailed theoretical treatment on the principles underlying these techniques.

Several of the experiments described in this chapter are based on pulsed (in some cases Fourier transform) techniques, similar to those routinely employed in NMR spectroscopy. Pulse EPR can, in many instances, offer significant advantages over CW methods, such as direct detection of relaxation times and access to longer distances between paramagnetic centres, through independent control of the electron and nuclear spins via application of short microwave (MW) and radiofrequency (RF) pulses. For a detailed description of the vector model and product operator formalism used in pulse techniques, the reader is referred to the two NMR textbooks in this series. The alternative operating modes of CW and pulse EPR are complementary methodologies; CW EPR is particularly powerful for identifying transitions *between* electron spin states, whereas the strength of pulsed EPR arises from observing transitions *within* electron spin states from electron–nuclear interactions.

See Hore et al., *NMR: The Toolkit* Oxford Chemistry Primer for more details on how pulse sequences work (see Bibliography).

9.2 Electron Nuclear DOuble Resonance (ENDOR) spectroscopy

Throughout this book, the experimental conditions used to observe the variety of EPR spectra have been obtained using a single irradiating MW frequency of low powers. In this situation, the spin populations among the Zeeman levels are not significantly perturbed compared to the thermal equilibrium values. However, in ENDOR spectroscopy two irradiating frequencies (MW and RF) of high power are simultaneously applied to the spin system. Electron and nuclear spin transitions are induced, and this is therefore referred to as a *double resonance experiment*. ENDOR can be performed in both CW and pulse mode, and the technique was first demonstrated by Feher in 1956.

In the ENDOR experiment, the NMR spectrum of the paramagnetic sample is indirectly detected through perturbations to the electron spin, resulting in a sensitivity enhancement of several orders of magnitude over NMR. The technique offers many advantages over EPR for the detailed structural characterization of paramagnetic systems in both solution and solid state, including the detection of very small hyperfine couplings and the precise identification of the nuclei responsible for the observed coupling.

CW ENDOR

The basic principles of CW ENDOR can be explained by reference to a simple two-spin system ($S = ½$, $I = ½$). The energy level diagram for this system was shown in Fig. 2.5, with the four possible energy levels represented as E_1–E_4. In this scheme, the two observed EPR transitions have difference frequencies, labelled EPR I ($E_2 \leftrightarrow E_4$ transition) and EPR II ($E_1 \leftrightarrow E_3$ transition), while the two allowed NMR transitions have difference frequencies labelled NMR I ($E_3 \leftrightarrow E_4$) and NMR II ($E_1 \leftrightarrow E_2$). Using low MW powers, resonance absorption occurs as, for example, the EPR II transition frequency $E_1 \leftrightarrow E_3$ is induced but rapid relaxation ensures that the excited spins return quickly to the ground state leading to an unsaturated EPR signal (Fig. 9.1a). Using higher MW powers, the $E_1 \leftrightarrow E_3$ transition becomes saturated such that the spin population in both levels equalizes (Fig. 9.1b). Under these conditions, the EPR signal intensity decreases as no further spin transitions occur. In order to restore the EPR signal intensity, a population difference between levels E_1 and E_3 must be created. One way to achieve this is by pumping the RF transition $E_3 \leftrightarrow E_4$ (NMR I) using a saturating RF field (Fig. 9.1c). An enhancement of the EPR absorption is detected if the irradiated RF frequency is resonant with this transition; this enhancement represents the first ENDOR signal.

A population difference between levels $E_1 \leftrightarrow E_3$ can alternatively be restored by pumping the second RF transition $E_1 \leftrightarrow E_2$, labelled NMR II (Fig. 9.1d). This creates an enhancement of the EPR signal, which represents the second ENDOR resonance line. Since the quanta of radiowaves required to induce transitions NMR I and NMR II

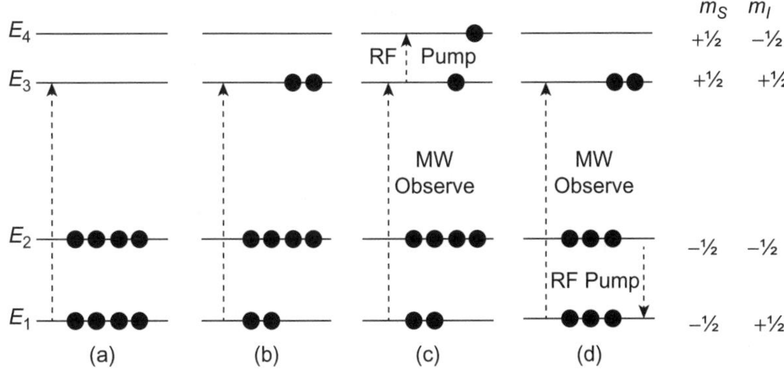

Fig. 9.1 Energy level diagram for the two-spin system $S = ½$, $I = ½$ (shown in Fig. 2.5), illustrating the effects of the applied MW and RF powers; (a) application of low MW power (to induce transition EPR II), (b) saturating MW power, (c) saturating RF power (to induce transition NMR I), and (d) saturating RF power (to induce transition NMR II).

are different, the two ENDOR resonances appear at different irradiated radiofrequencies in the spectrum. Therefore, in an ENDOR experiment, the nuclear transitions are detected via observation of the desaturation of the EPR transition as a function of the applied radiofrequency.

The allowed EPR and NMR transition frequencies for the two-spin $S = I = ½$ system are shown again in Fig. 9.2(a), labelled $\nu_{e1,2}$ and $\nu_{N1,2}$. In an isotropic EPR spectrum the resonances EPR I and EPR II appear as two lines at different magnetic field positions separated by a (Fig. 9.2b). In the resulting ENDOR spectrum, the two NMR transitions occur at the frequencies:

$$\nu_N = \left|\nu_L \pm \frac{a}{2}\right| \tag{9.1}$$

The narrow field separation between EPR I and EPR II for small a values results in only a negligible shift in ν_L.

where ν_L is the nuclear Larmor frequency of the nucleus contributing to the nuclear Zeeman levels. For spin systems where $\nu_L > |a|/2$, referred to as the weak coupling regime, the NMR transitions in the ENDOR spectrum are centred at ν_L and separated by $|a|$ (Fig. 9.2c). This situation is typical of protons, particularly at high fields. The same two ENDOR lines are observed when either EPR transition is observed (i.e. EPR I or EPR II). However, because ν_L is magnetic field dependent, the centre of the ENDOR lines observed by saturating EPR I or EPR II will be slightly shifted in frequency.

Weak coupling regime: $\nu_L > |a/2|$
Strong coupling regime: $\nu_L < |a/2|$.

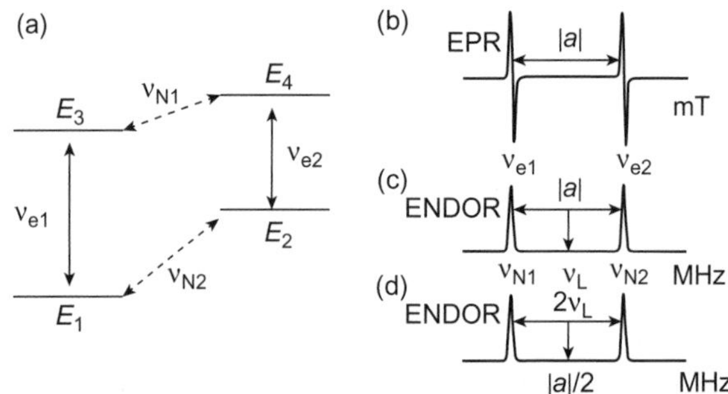

Fig. 9.2 (a) The energy levels for the two-spin system $S = ½$, $I = ½$, indicating the allowed EPR (solid) and NMR (dashed) transitions; (b) resonances ν_{e1} and ν_{e2} appear at different magnetic field positions in the EPR spectrum. The ENDOR resonances ν_{N1} and ν_{N2} in the (c) weak coupling and (d) strong coupling regimes.

Table 9.1 Nuclear Larmor frequencies at 350 mT (B_{ref}).

Nucleus	ν_L (in MHz) @ 350 mT
^{1}H	14.90212
^{2}H	2.28757
^{14}N	1.077200
^{15}N	1.511040
^{17}O	2.02098
^{31}P	6.03801

Alternatively, in some cases $\nu_L < |a|/2$. This is referred to as the *strong coupling* regime which occurs for ^{2}H, ^{13}C, and ^{14}N particularly at low field. In this case, the two ENDOR lines are separated by $2\nu_L$ and centred at $|a|/2$ (Fig. 9.2d). In this regime, the separation ($2\nu_L$) between the ENDOR resonances will be slightly different when either EPR line (i.e. EPR I or EPR II) is monitored owing to the magnetic field dependency of ν_L. The Larmor frequency at the selected field position (ν_B) in the ENDOR experiment, can be calculated according to:

$$\nu_B = \frac{\nu_{L(ref)}B}{B_{ref}} \tag{9.2}$$

where $\nu_{L(ref)}$ is the Larmor frequency determined at a reference field, B_{ref} (Table 9.1). Therefore, in an ENDOR experiment the hyperfine coupling constant (a) can be measured with high resolution and accuracy and can also be directly assigned to a specific nucleus since the values of ν_L are specific to individual nuclei.

Resolution enhancement

One important advantage of ENDOR is the resolution enhancement gained for organic radicals in solution. As seen in Chapter 4, the EPR spectra of organic radicals can become very complicated even for a small number of nuclei with $I > ½$, due to the fact that addition of non-equivalent nuclei to the paramagnetic system causes a *multiplicative* increase in the number of lines in the EPR spectrum. In contrast, each group of equivalent nuclei (of $I = ½$) contributes only a pair of lines to the isotropic ENDOR spectrum, regardless of the number of inequivalent nuclei. The resulting ENDOR spectra are therefore considerably simplified due to the *additive* increase in the number of lines. For example, the solution EPR spectrum of the benzyl radical should produce fifty-four lines (see Chapter 4). By comparison, the ENDOR spectrum produces only four pairs of lines (Fig. 9.3), from which the hyperfine couplings can be determined directly. Notice that three of the proton environments in the benzyl radical give rise to signals in the weak coupling regime (dotted arrows), with one resonance in the strong coupling regime (dashed arrow).

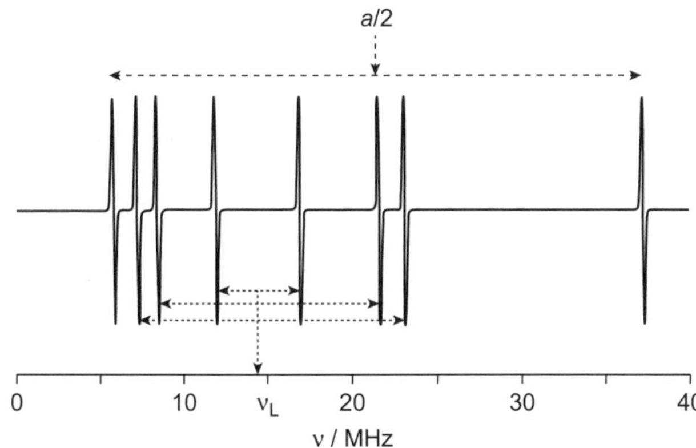

Fig. 9.3 Isotropic ^{1}H ENDOR spectrum of the benzyl radical indicating the resolution enhancement over the corresponding EPR spectrum (cf. Fig. 4.12). Three ^{1}H environments lead to resonances in the weak coupling regime (centred on ν_L), and one environment leads to ENDOR lines in the strong coupling regime (centred on $a/2$).

As discussed in Chapter 6, the EPR spectra of transition metal ions and their complexes typically display anisotropic **g** and **A** tensors. In order to extract the full anisotropic hyperfine tensor arising from delocalization of the unpaired electron onto surrounding ligands nuclei, *orientation-selective* hyperfine measurements should be performed (section 9.6). Analysis of the complete hyperfine tensor enables one to determine the isotropic (Fermi contact) and anisotropic (dipolar) contributions, which can yield structural information on the complex (see Chapter 5). The transition frequencies $\nu_\pm$ of the observed anisotropic hyperfine resonances are calculated by:

$$\nu_\pm = \left[\sum_{i=1}^{3} \left[\frac{m_S}{g(\theta,\phi)} \left(\sum_{j=1}^{3} g_j l_j A_{ji} \right) - l_i \nu_N \right]^2 \right]^{1/2} \tag{9.3}$$

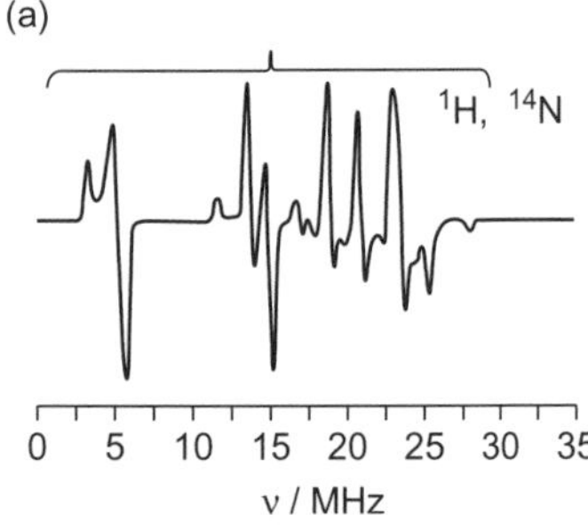

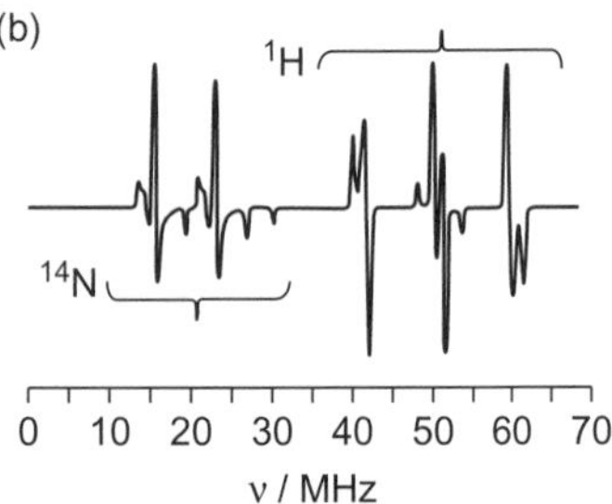

Fig. 9.4 The frozen solution ^{1}H and ^{14}N ENDOR spectra of the [Cu^{II}(salen)] complex recorded at (a) 9 and (b) 35 GHz (see Fig. 6.13 for the X-band EPR spectrum).

After S. Kita, M. Hashimoto, and M. Iwaizumi, *Inorg. Chem.* 1979, **18**, 3432.

In pulse techniques, short MW or RF pulses rotate the magnetization from thermal equilibrium by an amount given by their flip angle (symbol β, not to be confused with the β-spin state).

Since time and frequency domain are related by Fourier transformation, *short* on the time scale relates to *broad* on the frequency scale. Hence, long pulses relate to narrow excitation bands (i.e. selective).

where A_{ji} is the orientation-dependent value of the hyperfine coupling and l_i are the direction cosines (see section 5.2).

The interpretation of X-band ENDOR spectra can be complicated in cases where resonances from weakly coupled ^{1}H and strongly coupled ^{14}N nuclei overlap, as demonstrated in Fig. 9.4. ENDOR measurements at higher microwave frequencies (e.g. Q-band) can sometimes eliminate this problem, since at higher fields these nuclei with different Larmor frequencies will be well separated. Additional benefits to high-field (or high-frequency) ENDOR include increased orientation selectivity, and increased sensitivity, which is particularly advantageous in studies of metalloenzymes when only a small amount of sample may be available.

CW ENDOR spectroscopy is hence an extremely powerful technique for the simplification of spectra arising from complex organic radicals and transition metal ions with unresolved hyperfine couplings from the surrounding nuclei. However, a significant disadvantage of CW ENDOR spectroscopy is the requirement to meet the saturation conditions for both the EPR and NMR transitions, which demands a critical balance between the rates of induced transitions and relaxation. More significantly, even when all experimental conditions have been optimized, the ENDOR signal accounts for only a small percentage of the change in EPR signal intensity. Fortunately, these shortcomings can be overcome using pulsed ENDOR methods.

Pulsed ENDOR

Pulsed ENDOR spectroscopy offers several advantages over CW ENDOR, in particular greater sensitivity, higher spectral resolution and (often) reduced measurement acquisition times. In contrast to CW ENDOR, during which the MW and RF fields are continuously applied to the sample, in a pulsed ENDOR experiment short pulses of microwaves (on the order of ns) and radio waves (timescale of μs) are used. The amplitude of the electron spin echo is then subsequently measured as a function of the applied radio frequency. The most significant benefit of pulsed ENDOR is the ability to independently select pulse lengths and delay times, which overcomes the requirement to balance relaxation and induced transition rates, thereby affording measurements across a wider temperature range than in CW mode. Furthermore, the two commonly used pulse sequences, Davies and Mims ENDOR, can be modified to selectively manipulate spins in order to simplify complicated spectra containing signals from multiple nuclei.

The Davies ENDOR sequence, written as $\pi_{MW} - \pi_{RF} - \pi/2_{MW} - \tau - \pi_{MW} - \tau -$ echo (Fig. 9.5), is a population-transfer experiment and can be considered as the pulsed equivalent of a CW ENDOR experiment. The effect of the pulses on the electron and nuclear spin energy level populations are illustrated in Fig. 9.6 for the two-spin system ($S = I = ½$) considered previously (Fig. 9.1). The first selective π_{MW} pulse inverts the electron spin populations in the E_1 and E_3 manifolds (Fig. 9.6a), i.e. induces the transition EPR II. This is followed by a π_{RF} pulse that inverts the nuclear spin populations upon resonance with an NMR transition (e.g. NMR I, Fig. 9.6b). The remaining

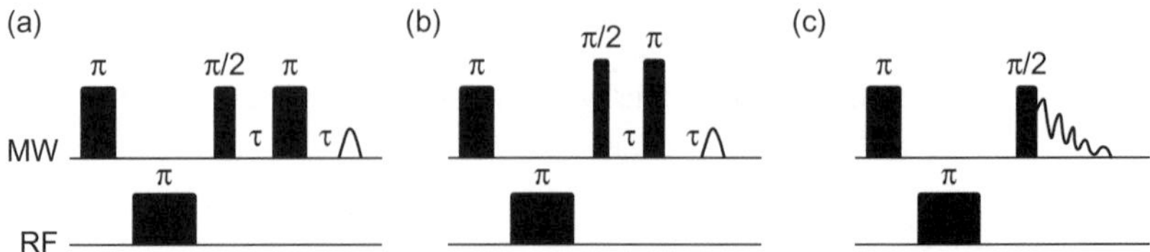

Fig. 9.5 The pulse sequences for echo detected Davies ENDOR using (a) selective, (b) non-selective, and (c) FID detection.

$\pi/2 - \tau - \pi - \tau -$ echo sequence consists of either the selective (Fig. 9.5a) or non-selective (Fig. 9.5b) electron-spin echo detection pulses.

Using non-selective detection pulses affords superior sensitivity as the delay time between the pulses can be reduced, which minimizes unwanted echo decay due to the phase memory time, T_M, an empirical parameter, which is affected by processes such as spin-spin relaxation and lifetime broadening (see Chapter 2). Alternatively, if T_M is very short, it is possible to replace the two-pulse echo detection sequence with a single selective $\pi/2_{MW}$ pulse to record the free-induction decay (Fig. 9.5c).

In order to optimize the pulsed ENDOR response, the pulse lengths and delays need to be carefully selected to ensure full inversion of the electron and nuclear spin energy level pairs. An important requirement of Davies ENDOR is that the first π_{MW} pulse only excites one of the allowed EPR transitions of each spin packet (i.e. $E_1 \rightarrow E_3$ (EPR II) or $E_2 \rightarrow E_4$ (EPR I)). Therefore, the inversion bandwidth ($\Delta\nu = 1/t_p$) of π_{MW} is related to the isotropic hyperfine splitting a_{iso} through the selectivity parameter, η_S:

$$\eta_S = \frac{a_{iso}\, t_{MW}}{2\pi} \tag{9.4}$$

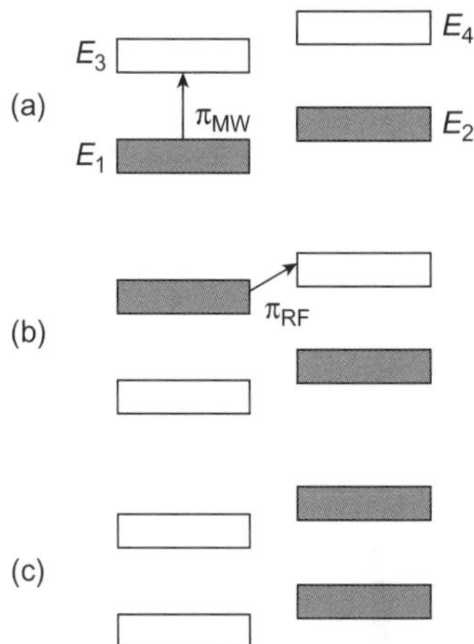

Fig. 9.6 (a) The π_{MW} pulse selectively inverts electron spin populations, (b) the π_{RF} pulse selectively inverts nuclear spin populations resulting in electron spin saturation (c).

Experimentally it is convenient to use the same field strength for the three MW pulses in the Davies ENDOR sequence (Fig. 9.5a). Therefore the pulse length, t_p, must be varied to achieve the different *tip angles* (e.g. for a fixed B_1 value a π-pulse is twice as long in duration as a $\pi/2$-pulse). The length of the inversion RF pulse, t_p, is given by:

$$t_p = \frac{\pi}{\gamma_N B_2} \tag{9.5}$$

where B_2 is the RF magnetic field strength. From eqn 9.5, nuclei with different gyromagnetic ratios will require different pulse lengths to achieve inversion.

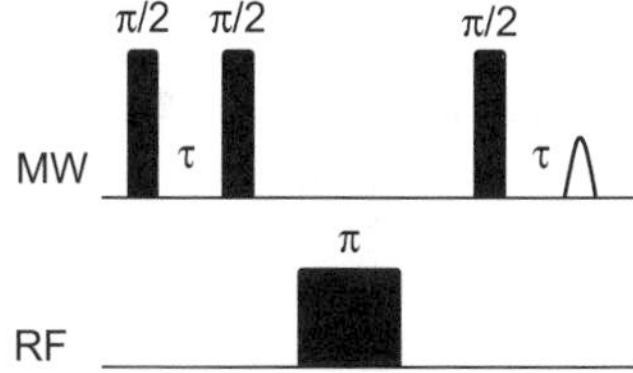

Fig. 9.7 The Mims ENDOR pulse sequence.

The alternative Mims ENDOR sequence, written as $\pi/2_{MW} - \tau - \pi/2_{MW} - \pi_{RF} - \pi/2_{MW} - \tau -$ echo (Fig. 9.7), is particularly effective for weakly coupled nuclei, whereas the Davies experiment would require a very long inversion pulse resulting in excitation of only very few spin packets. Mims ENDOR is based on a stimulated echo sequence with three non-selective $\pi/2_{MW}$ pulses. The first two preparation MW pulses generate an oscillatory polarization pattern in the EPR line with a period $1/\tau$. The third detection MW pulse generates a stimulated echo from this oscillatory pattern. When the π_{RF} pulse is resonant with one of the nuclear transitions, a quarter of this pattern is shifted by $+a_{iso}$ and another quarter is shifted by $-a_{iso}$ along the EPR frequency axis. For $a_{iso}\tau = \pi$, this leads to complete destructive interference of the oscillatory pattern and the stimulated echo is cancelled. For $a_{iso}\tau = 2k\pi$ with integer k, the pattern is virtually unchanged and the stimulated echo persists. The ENDOR efficiency (F_{ENDOR}) in a Mims experiment thus has a cosine dependency on the magnitude of the observed hyperfine coupling, a_{iso}, and the interpulse time delay, τ, given by:

Polarization is magnetization aligned along the *z* axis (i.e. the quantization axis).

$$F_{ENDOR} = \frac{1}{4}\left(1 - \cos(a_{iso}\tau)\right) \tag{9.6}$$

Therefore, at certain combinations of a_{iso} and τ, the ENDOR response reduces to zero; this is referred to as the *blind-spot* behaviour. In order to avoid loss of spectral information, it is necessary to add the results of several Mims experiments recorded with different τ values.

Although pulse ENDOR is less susceptible to instrumental artefacts than CW methods (due to the absence of applied MW and RF fields during detection), asymmetric line intensities for the low and high frequency components of a hyperfine splitting are sometimes observed in pulse ENDOR due to the non-linearity of the B_2 field as a function of frequency or due to slow nuclear spin relaxation.

9.3 Two-pulse and three-pulse Electron Spin Echo Envelope Modulation (ESEEM)

Whilst CW and pulsed ENDOR spectroscopy are experiments performed directly in the frequency domain, alternative time-domain experiments such as ESEEM spectroscopy are available that can also be used to extract weak ligand hyperfine couplings. ESEEM provides complementary information to ENDOR spectroscopy, and is particularly powerful for detecting low frequency (< 5 MHz) couplings. ESEEM monitors the NMR frequencies indirectly through observation of the mixing of allowed and (formally) forbidden EPR transitions as modulations superimposed on a time-decaying spin echo. Echo envelope modulation occurs when state mixing of hyperfine levels occurs. Therefore strong ESEEM is generally observed either when the hyperfine interaction is comparable to the nuclear Zeeman interaction and has a significant anisotropic contribution, or when the nuclear quadrupolar interaction is comparable to the hyperfine (e.g. remote ^{14}N couplings).

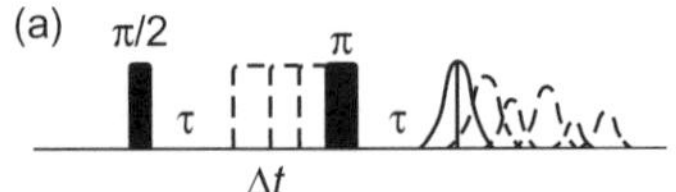

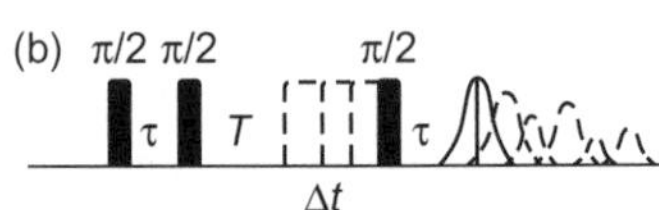

Fig. 9.8 (a) The two-pulse ESEEM sequence, in which the time interval τ is incremented by the dwell time Δt; a primary echo is observed at time τ after the second pulse. (b) The three-pulse ESEEM sequence, in which τ is kept constant and the time interval T is incremented.

In the two-pulse ESEEM experiment, a simple $\pi/2 - \tau - \pi$ pulse sequence generates a primary echo detected at time τ after the second pulse (Fig. 9.8a). As the inter-pulse delay time τ is incremented, the echo intensity decays with the phase memory time, T_M, and is modulated by the nuclear transition frequencies ν_{N1} and ν_{N2} (also labelled as angular frequencies ω_{34} and ω_{12}) originating in each electron spin manifold. In addition to the principal hyperfine frequencies, in two-pulse ESEEM the sum and difference combination frequencies ($\omega_+ = |\omega_{12} + \omega_{34}|$ and $\omega_- = |\omega_{12} - \omega_{34}|$; Fig. 9.9) also contribute to the resulting modulation, $V_{2\text{-pulse}}$, which is described by:

$$V_{2\text{-pulse}}(\tau) = 1 - \frac{k}{4}\left[2 - 2\cos(\omega_{34}\tau) - 2\cos(\omega_{12}\tau) + \cos(\omega_+\tau) + \cos(\omega_-\tau)\right] \tag{9.7}$$

where the modulation depth parameter, k, of the oscillations superimposed on an exponential decay is defined as:

In this eqn, r refers to the electron-nucleus distance.

$$k = \frac{9}{4}\left(\frac{\mu_0 g \mu_B}{4\pi B}\right)^2 \frac{\sin^2(2\theta)}{r^6} \tag{9.8}$$

The modulations of the electron spin echo due to couplings with j nuclei combine in a multiplicative manner, and are superimposed on the exponential decay of the spin echo, resulting in the modulated spin echo time trace:

$$V'_{2\text{-pulse}}(\tau) = \exp\left(\frac{-2\tau}{T_M}\right)\prod_j V_{2\text{-pulse},j}(\tau) \tag{9.9}$$

During the spectrometer deadtime, spin echoes cannot be recorded in order to prevent damage to the detector through overloading.

Fourier transformation of the resulting modulated time trace results in spectra in the frequency domain containing the nuclear frequencies (Fig. 9.9). As seen from eqn 9.9, the modulated time trace is a function of the phase memory time, T_M, which is typically short. This presents a major limitation to the two-pulse ESEEM experiment because low frequency modulations can sometimes be unresolved. Another potential source of information loss is created by the spectrometer deadtime, τ_d (~100 ns at X-band frequency). This can lead to distortions or artefacts in the resulting frequency-domain spectrum, although the information lost during the

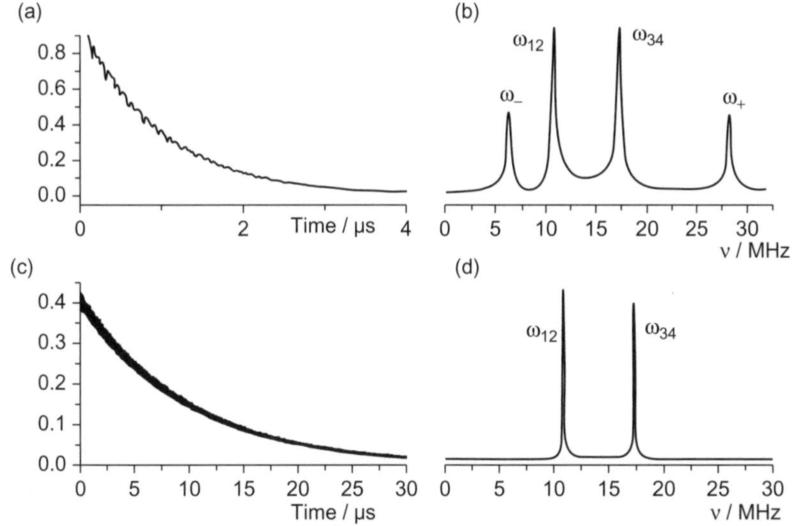

Fig. 9.9 (a) Standard two-pulse ESEEM, $\pi/2 - \tau - \pi - \tau$ – echo; (b) the corresponding frequency-domain spectrum obtained after Fourier transformation; (c) Standard three-pulse ESEEM, $\pi/2 - \tau - \pi/2 - \tau - \pi/2 - \tau$ – echo; (d) the corresponding frequency-domain spectrum. Combination peaks are only visible in the two-pulse spectra (absolute magnitude shown).

$I = ½$ nuclei with $a >> \nu_L$ do not give noticeable ESEEM modulation.

The most favourable situation for observation of narrow quadrupole lines from ^{14}N nuclei, called the cancellation condition, occurs when $\nu_L = |½a|$.

deadtime can in principle be recovered through the use of remote-echo detection schemes.

Some of the limitations of two-pulse ESEEM can be overcome through the use of the three-pulse ESEEM scheme (Fig. 9.8b). The overall sequence, written as $\pi/2 - \tau - \pi/2 - T - \pi/2$, generates a *stimulated* echo observed at time τ after the third microwave pulse. A constant τ value is used, whilst the time delay T between the second and third $\pi/2$ pulses is incremented. This pulse sequence results in an echo envelope that decays with the transverse nuclear relaxation time, modulated by the nuclear transition frequencies, given by:

A stimulated echo occurs from the action of three or more MW pulses. The decay of a stimulated echo is a function of T_1.

$$V_{3\text{-pulse}}(\tau, T) = 1 - \frac{k}{4}\{[1-\cos(\omega_{12}\tau)][1-\cos(\omega_{34}(\tau+T))] + [1-\cos(\omega_{34}\tau)][1-\cos(\omega_{12}(\tau+T))]\} \quad (9.10)$$

The experimental time trace of an electron coupled to a single nuclear spin is given by:

$$V'_{3\text{-pulse}}(\tau, T) = \exp\left(\frac{-T}{T_M^{(N)}}\right)\exp\left(\frac{-2\tau}{T_M}\right)V_{3\text{-pulse}}(\tau, T) \quad (9.11)$$

where $T_M^{(N)}$ is the nuclear phase memory time (which is longer than T_M), leading to narrower lines in the frequency-domain spectrum and therefore providing increased spectral resolution. The appearance of three-pulse ESEEM spectra are considerably simplified in comparison to the analogous two-pulse experiment due to the absence of the sum and difference combination frequencies. However, from the τ-dependent cosine terms in eqn 9.10, three-pulse ESEEM is also affected by blind-spot behaviour (cf. Mims ENDOR).

9.4 Hyperfine Sublevel Correlation (HYSCORE) spectroscopy

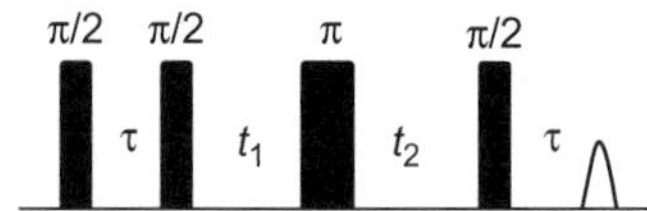

Fig. 9.10 The HYSCORE sequence, which has an additional π-pulse inserted into the three-pulse ESEEM experiment (cf. Fig. 9.8(b)).

Nuclear coherence is magnetization perpendicular to the quantization axis.

The HYSCORE (*HYperfine Sublevel CORrElation*) pulse sequence is a four-pulse MW sequence in which an additional mixing π pulse is inserted between the second and third π/2 pulse of the three-pulse ESEEM experiment (Fig. 9.10). The two inter-pulse delays, t_1 and t_2, are varied independently to produce a two-dimensional (2D) time delay array. The nuclear coherence generated by the first two π/2 pulses undergoes free evolution during time t_1 with frequency ω_{12} (ω_{34}). The mixing π pulse then transfers populations in one m_S manifold to the other and similarly transfers the nuclear coherence between manifolds so that it evolves with frequency ω_{34} (ω_{12}) during time t_2. The modulated time decay data is subsequently Fourier transformed in both dimensions (i.e. t_1 and t_2) to produce a 2D frequency-domain spectrum (with axes ν_1 and ν_2). The nuclear frequencies from the different m_S manifolds are correlated and appear as cross-peaks at the frequencies (ν_1, ν_2), (ν_2, ν_1), and (ν_1, $-\nu_2$), (ν_2, $-\nu_1$) in the (+, +) and (+, −) quadrants of the 2D spectrum, respectively. As strong cross peaks can only be observed between NMR frequencies of the same nucleus, HYSCORE spectra can be significantly simplified compared to three-pulse ESEEM.

Another advantage of HYSCORE spectroscopy is that the frequencies from weakly-coupled nuclei ($|a_{iso}| < 2|\nu_L|$) appear as cross-peaks in the (+, +) quadrant, whereas strongly-coupled nuclei ($|a_{iso}| > 2|\nu_L|$) are observed in the (+, −) quadrant (Fig. 9.11a). This facilitates spectral interpretation for systems containing many interacting nuclei such as metalloenzymes or proteins. For disordered systems with broad ESEEM features, the correlation peaks broaden into ridges, as illustrated for the two-spin system ($S = I = ½$) with an axial hyperfine tensor (Fig. 9.11b). The anisotropy of the dipolar hyperfine interaction, T, can be determined from the maximum curvature of the ridges away from the anti-diagonal, labelled ω_{max} ($= 9T^2/32|\omega_L|$) and the magnitude of a_{iso} can be found from the ridge end points or from simulation.

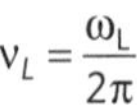

$$\nu_L = \frac{\omega_L}{2\pi}$$

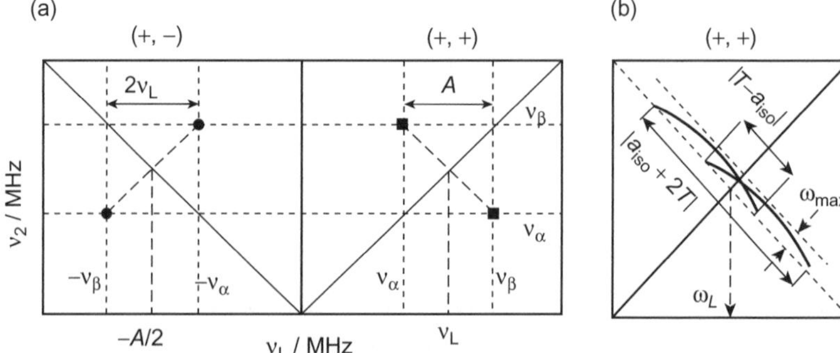

Fig. 9.11 (a) Two-dimensional HYSCORE spectrum where full squares ■ represent cross peaks from weakly coupled nuclei in the (+, +) quadrant, and full circles ● represent cross peaks from strongly coupled nuclei in the (−, +) quadrant. ν_L is the Larmor frequency for the nucleus of interest, A is the hyperfine coupling, ν_α ($= \omega_{12}$) and ν_β ($= \omega_{34}$); (b) (+, +) quadrant for the powder HYSCORE pattern for an $S = I = ½$ spin system with an axial hyperfine tensor.

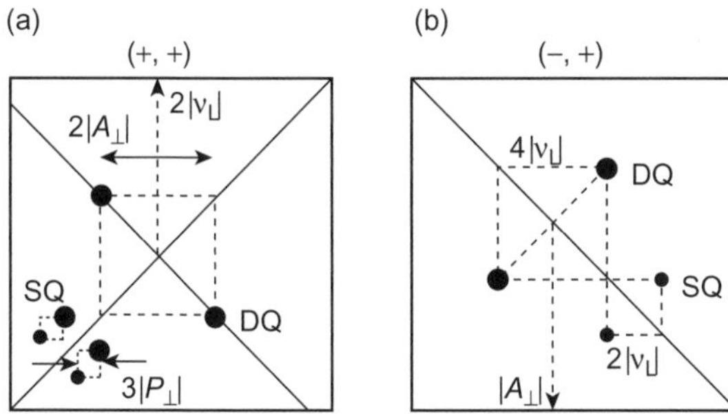

Fig. 9.12 HYSCORE spectra for a spin system with one $I = 1$ nucleus and non-zero quadrupole interaction for (a) weak and (b) strong hyperfine interactions.

HYSCORE spectroscopy is particularly sensitive to low-frequency signals (<5 MHz), and is therefore an ideal tool to study weakly coupled nitrogen nuclei (I = 1). The resulting correlation patterns are considerably more complicated due to single quantum (SQ: $|\Delta m_I| = 1$) and double quantum (DQ: $|\Delta m_I| = 2$) transitions (Fig. 9.12). The DQ transitions are additionally influenced by the nuclear quadrupole contribution, and are observed at frequencies V^{DQ} given by:

Double quantum transitions:

ν_{β}^{DQ}: $\left(m_S = -\frac{1}{2}, m_I = 1\right) \rightarrow (m_S = -\frac{1}{2}, m_I = -1)$;

ν_{α}^{DQ}: $\left(m_S = \frac{1}{2}, m_I = -1\right) \rightarrow (m_S = \frac{1}{2}, m_I = 1)$.

$$V_{\alpha,\beta}^{DQ} = 2\sqrt{\left(\nu_I \pm \frac{a}{2}\right)^2 + \left(\frac{e^2qQ}{4h}\right)^2 (3+\eta^2)} \tag{9.12}$$

where a is the hyperfine coupling, Q is the electric quadrupole moment, and η is the asymmetry parameter (see section 5.5).

The HYSCORE experiment involves detection of a stimulated echo, and is hence affected by τ-dependent blind spots, as described above for three-pulse ESEEM. The *Nyquist* frequency (or maximum detectable frequency) is half of the inverse of the dwell time, Δt, which is defined as the time increment of t_1 (and t_2) between each pulse cycle. The excitation range of the mixing π pulse is an important variable in HYSCORE spectroscopy. It is common that the π pulse is not able to fully transfer all of the nuclear coherence between the two m_S manifolds. Therefore packets with frequency ω_{12} before the π pulse will continue to evolve with this frequency after the pulse; this leads to peaks on the diagonal, which can cause resolution problems, in particular for weakly coupled nuclei with large gyromagnetic ratios.

9.5 **Pulsed Electron DOuble Resonance (PELDOR) spectroscopy**

A number of techniques are now available that facilitate the determination of electron-electron or electron-nuclear distances, all involving analysis of the dipolar interaction (see Chapter 7). CW measurements used to characterize the interaction between spins typically yield measurements of distances up to 2–2.5 nm (due to the dependence on T_1, T_2, and diffusion times), and rely on detailed lineshape analysis (see Chapter 7). In contrast, pulsed EPR techniques can access distances in the range 1.8–6 nm (even up to 10 .2 nm using deuterated soluble proteins). A pulsed EPR technique used for distance determination is PELDOR, also termed DEER (*Double Electron-Electron Resonance*). This experiment measures the dipolar interaction between two paramagnetic centres

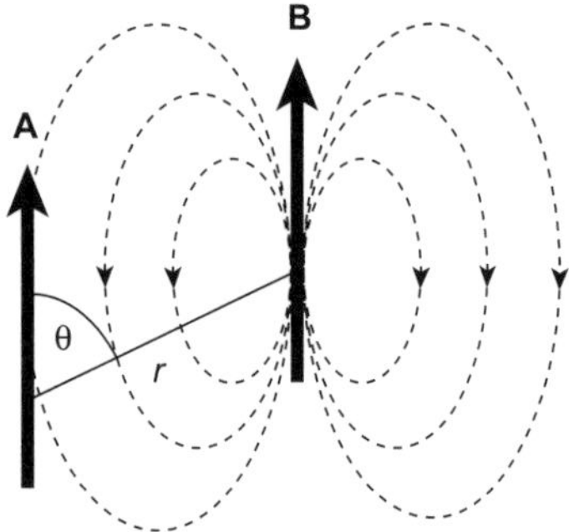

Fig. 9.13 The classical dipolar interaction between two magnetic dipoles, in this case two electron spins, labelled A and B, is a function of the separation distance, r, and their relative orientation, defined by θ (cf. Fig. 5.13).

Na^+

Gd(III)-4-vinyl-PyMTA.

(labelled A and B in Fig. 9.13). The technique is frequently used to study membrane dynamics and to elucidate protein structures, which can be difficult to achieve via X-ray crystallography due to the challenges associated with obtaining good quality single crystals of high molecular weight biomolecules. The paramagnetic centres can be intrinsic metal ions, or artificially incorporated spin labels that are attached through site-directed spin labelling, such as nitroxides for protein structure determination (e.g. MTSSL, see Chapter 8) or Gd(III) centres for in-cell applications (e.g. Gd-4-vinyl-PyMTA).

The four-pulse DEER sequence uses two MW sources of different frequency, referred to as the observer (ν_1) and pump (ν_2) pulses. The frequency offset, $\nu_{offset} = \nu_2 - \nu_1$, should ensure that the pump pulse does not directly excite the observer spins (Fig. 9.14a). The first two observer pulses therefore create an echo of the spin packet at the observer frequency ν_1 (labelled 'A spins'), and the pump pulse (ν_2) inverts the spin magnetic moments of a different resonance spin packet (labelled 'B spins'). The third observer pulse creates a refocused echo at time $t = 2\tau_1 + 2\tau_2$. When nitroxide spin labels are used at X-band, ν_1 is often selected as the low-field component of the nitroxide radical signal, exciting only a small number of spins that have an alignment of the z axis parallel

Orientation selectivity in X-band PELDOR is achieved by incrementing ν_{offset} to acquire data sets where orientations of ***B*** close to the x,y axes of the >N–O• spin label are detected.

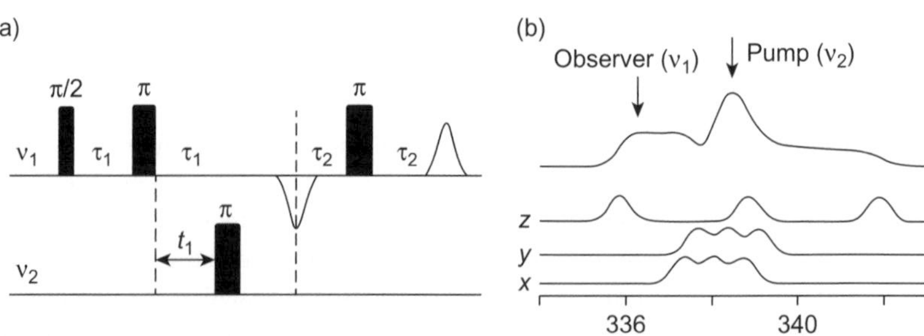

Fig. 9.14 (a) The four-pulse DEER sequence applies observer (ν_1) and pump (ν_2) pulses at different microwave frequencies. Delays τ_1 and τ_2 are held constant, while delay t_1 is incremented. (b) Corresponding summed and decomposed individual EPR spectra.

to B (Fig. 9.14b). To maximize the modulation depth, the pump pulse ν_2 is set to a field position corresponding to the most intense component of the nitroxide radical signal, which selects spins with all possible orientations of the nitroxide with respect to $\mathbf{B}$. At X-band, this results in ν_{offset} of ~70 MHz. Similar to high-frequency ENDOR, performing PELDOR at high microwave frequencies (e.g. 180 GHz) offers increased orientation-selection and information about the relative orientation of the spin labels, but suffers from inherently lower sensitivity due to a smaller fraction of excited spins.

ν_2 inverts the state of the B spin and hence changes the resonance frequency of the A spin due to dipole-dipole interaction. This results in the echo signal being modulated by the dipolar frequency ω_{AB}.

As the position of the pump pulse is incremented between the two π observer pulses, the dipolar coupling between the A and B spins modulates the echo intensity (I) by the electron-electron coupling (ω_{AB}), as described by:

$$I(t) = I_0 \cos\left(\omega_{AB}(t_1 - \tau_1)\right) \tag{9.13}$$

where the oscillation frequency between the spins is given by the relation:

$$\omega_{AB} = \omega_{dd} + J = \frac{\mu_0 g_A g_B \mu_B^2}{2hr^3}\left(3\cos^2\theta - 1\right) + J \tag{9.14}$$

Here ω_{AB} is the dipole-dipole coupling, J is the exchange interaction (see Chapter 7), r is the distance between the spins, θ is the angle between $\mathbf{B}$ and the vector connecting the spins, and g_A, g_B are the g-values of the coupled spins. In the absence of orientation selection, Fourier transformation of the resulting echo modulation (i.e. the PELDOR time trace) yields the distribution of dipole-dipole coupling frequencies. For large distances ($r > 2$ nm) with narrow distributions, and negligible exchange interaction, the result is a Pake pattern (see sections 5.4 and 7.6). More sophisticated data analysis by *Tikhonov regularization* can yield the distribution of distances (Fig. 9.15). In general, a larger distance between spins A and B extends the period of the modulated spin echo decay. A primary limiting factor in PELDOR spectroscopy is the requirement to observe at least three periods of complete dipolar oscillation, which requires long phase memory times, T_M. In PELDOR experiments using nitroxides and proton-containing solvents or macromolecules, T_M values of ~2 μs are typical, which limits the observable inter-spin distance to ~6–8 nm.

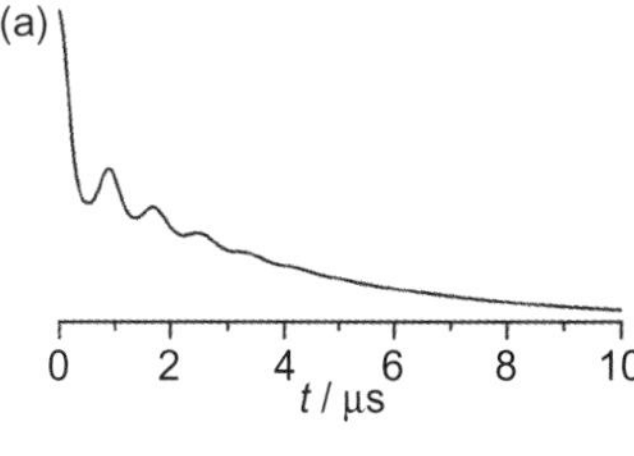

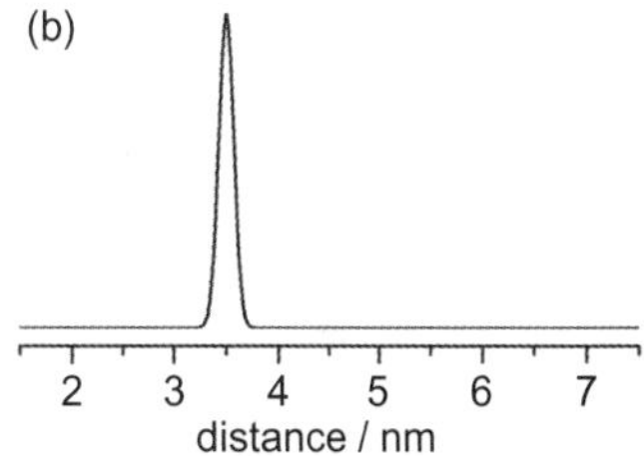

Fig. 9.15 (a) Modulated echo decay in a PELDOR experiment and (b) distance distribution following *Tikhonov regularization*.

9.6 Orientation selective hyperfine measurements

All of the advanced hyperfine measurements described in sections 9.2–9.5 are typically recorded at low temperatures, and therefore produce anisotropic EPR spectra (see Chapters 5 and 6). The elements of the hyperfine **A** tensor, and its relative orientation with respect to the **g** tensor, can be determined from variations in the magnitude of the hyperfine coupling at different magnetic field settings through *orientation selective* hyperfine measurements.

A polycrystalline EPR spectrum can be considered as a superposition of resonances from randomly orientated molecules, so that as the applied field B is swept it adopts all possible orientations with respect to the chosen molecular frame (Chapter 5). In orientation selective measurements (in particular for ENDOR), a series of hyperfine measurements are performed at a number of fixed magnetic field positions across the full width of the EPR spectrum. The hyperfine response at each field position arises only from the subset of molecules having orientations that contribute to the EPR intensity at

High-frequency measurements enable increased g-value resolution and are useful for orientation selective hyperfine spectroscopy.

that particular B field. The EPR resonance field positions, B_{res}, for given orientations (defined in terms of θ, ϕ) are expressed (to first order) as:

$$B_{res} = \left[\frac{h\nu - m_I A(\theta,\phi)}{\mu_B g(\theta,\phi)} \right] \tag{9.15}$$

where $A(\theta, \phi)$ and $g(\theta, \phi)$ may be obtained from eqns 9.16–9.17 as follows:

Equation 9.16 is analogous to eqn 9.3, used to determine the ENDOR resonance frequencies where l refers to the direction cosines (see Chapter 5).

$$A(\theta,\phi) = \left[\sum_{i=1}^{3} \left[\left(\sum_{j=1}^{3} g_j l_j A_{ji} \right) - l_i \nu_N \right]^2 \right]^{1/2} g(\theta,\phi)^{-1} \tag{9.16}$$

$$g(\theta,\phi) = \left[\sum_{i=1}^{3} (g_i l_i)^2 \right]^{1/2} \tag{9.17}$$

Determination of the anisotropic hyperfine tensor

The analysis of the hyperfine spectra recorded at different field positions can be used to determine the anisotropic **A** tensor, and hence the direction between an unpaired electron and an interacting nucleus of $I \neq 0$, through magnetic angle selection. To demonstrate the utility of orientation selective hyperfine measurements for the determination of anisotropic A values, consider the case of a simple $S = ½$ spin system in axial symmetry and interacting with a remote nucleus (e.g. a ^{1}H, $I = ½$) producing small A values which will be buried under the intrinsic EPR linewidth (and so not visible by CW EPR). The EPR spectrum and corresponding angular variation in resonant field is shown in Fig. 9.16 (analogous to Fig. 5.7a). Orientation selective hyperfine measurements should be performed at the unique *single-crystal* like positions (see section 5.3) labelled *i* and *iii*, and additionally at different field positions between these extremes (e.g. position *ii*). During this discussion, it is helpful to consider the paramagnetic centre to be fixed and treat the magnetic field as a vector within the **g** frame (as illustrated in Fig. 5.2).

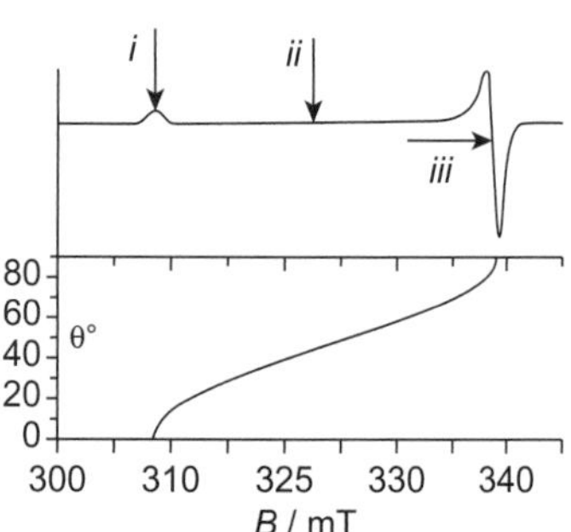

Fig. 9.16 EPR spectrum and angular variation in resonant field positions for an $S = ½$ spin system in axial symmetry.

In this axial system, at the magnetic field corresponding to $g_{\|}$ or $g_{\perp}$, the angle θ equals 0° or 90°, respectively. At all other B_{res} positions between these two extremes the exact value of θ and hence the angular curve in Fig. 5.16, may be calculated as described in section 5.3. For rhombic systems, the angle ϕ must also be considered in the angular dependency curve (see Fig. 5.7b).

Orientation selective hyperfine measurements should be taken at different fixed field positions across the entire magnetic field range of the EPR spectrum (e.g. positions marked *i* to *iii* in Fig. 9.16).

The procedure for determining the anisotropic A values for the remote interacting nucleus will be illustrated here for two cases: i) where **g** (solid lines) and **A** (dashed lines) have coincident axes that are aligned, (Fig. 9.17a) and ii) where **g** and **A** have coincident axes that are 90° to each other (Fig. 9.17b). In Fig. 9.17a, the principal axes of **g** and **A** are aligned, so when B_{res} corresponds to $B_{\|}$ (the resonance field position *i* for $g_{\|}$, Fig. 9.16), the resulting hyperfine measurement will contain only one set of hyperfine lines separated by the $A_{\|}$ component (top spectrum in Fig. 9.17a). Similarly, when B_{res} corresponds to $B_{\perp}$ (the $g_{\perp}$ position *iii*), the hyperfine lines will be separated by the $A_{\perp}$ component (bottom spectrum in Fig. 9.17a).

At all intermediate B_{res} positions (e.g. position *ii*), the hyperfine lines are separated by an intermediate A value, calculated from eqns 9.15–9.17. Hence, the $A_{\|}$ and $A_{\perp}$ components of the hyperfine can be easily determined, in this case from the data recorded exclusively at positions *i* and *iii* only.

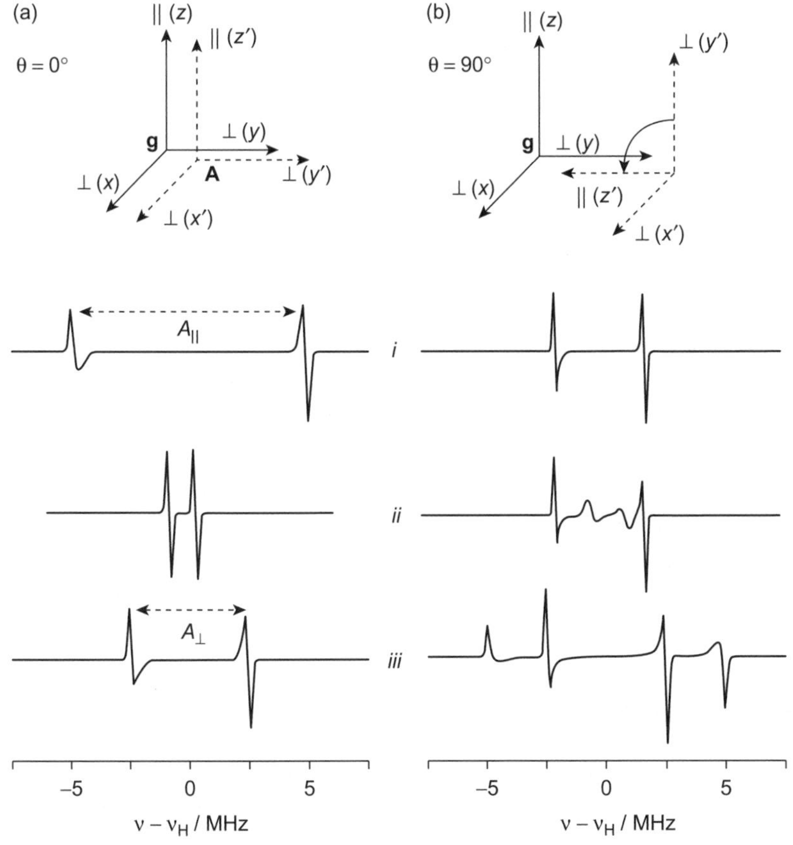

Fig. 9.17 Magnetic field orientation dependence of hyperfine measurements for a coordinate **g** frame (labelled *x*, *y*, *z*) and **A** frame (labelled *x′*, *y′*, *z′*), in which the respective coincident axes are aligned (a), with $\theta = 0°$, and orthogonal (b) with $\theta = 90°$.

A different orientation-selective hyperfine profile will be observed when the principal **g** and **A** axes are coincident but orthogonal to each other (Fig. 9.17b). In this case, the $g_{\|}$ and $A_{\|}$ components are 90° to each other, so that the parallel-axis of **g** is aligned with the perpendicular-axis of **A**. Hence when B_{res} corresponds to $g_{\|}$, the $A_{\perp}$ component of the hyperfine is observed (top spectrum, Fig. 9.17b). Alternatively, when the static magnetic field position corresponds to $g_{\perp}$, both components of the dipolar hyperfine are selected, producing two sets of lines separated by $A_{\|}$ and $A_{\perp}$. The $A_{\perp}$ coupling is observed at all magnetic field positions across the EPR spectrum. At the intermediate field position (i.e. position *ii*), the remaining hyperfine coupling can again be calculated from eqn 9.15. Hence, the values of $A_{\|}$, $A_{\perp}$, and θ can be found from the hyperfine measurement recorded at $B_{res} = g_{\|}$.

Finally, when the **g** and **A** axes are completely non-coincident, the two tensor principal axes systems must be related to each other by three Euler angles, and the hyperfine spectra at all B_{res} positions are considerably more complicated than for the two cases described above. The procedure for determining the angles of non-coincidence and identifying the principal components of, for example, the **A** tensor, is now considerably more involved, requiring matrix transformation (see Chapter 5).

9.7 Summary

- A number of advanced EPR techniques are available which considerably enhance the amount of information that can be extracted from a paramagnetic system, compared to CW EPR alone.
- In ENDOR spectroscopy, very small hyperfine interactions can be detected in both solution and solid state. The solution phase ENDOR spectrum leads to considerable simplification of the CW EPR spectrum.
- In ESEEM spectroscopy, low-frequency hyperfine interactions can be measured very accurately. The ESEEM measurement can be affected by combination peaks (two-pulse) and τ-dependent blind spots (three-pulse).
- In HYSCORE spectroscopy, resonances from strongly coupled and weakly coupled nuclei appear in different quadrants. The technique is very effective for analysing quadrupole interactions.
- In PELDOR (or DEER) spectroscopy, distance distributions between electron spins can be obtained.
- Orientation selective measurements, in ENDOR, ESEEM, HYSCORE, and PELDOR, are necessary to understand the relationship between the **g** anisotropy and the dipolar interaction under investigation.

9.8 Exercises

9.1) Given that the precession angular frequency for a charged particle in a static magnetic field is $\omega = \gamma B$, calculate the Larmor frequency for an electron and a proton in a field of 1.5 T.

9.2) For a B_2 RF field of 10 mT, calculate the pulse duration time required to produce a flip angle of 90° and 180° for a proton.

9.3) Calculate the number of lines expected in the EPR and ENDOR spectra of the phenalenyl radical assuming no superposition of overlapping lines (see Fig. 4.22 for radical structure).

9.4) Determine the values of $A_{\parallel}$ and $A_{\perp}$ in Fig. 9.17(b) and hence sketch the profile of the expected powder hyperfine spectrum for the spin system shown in Fig. 9.16 assuming a quasi-isotropic **g** tensor.

9.5) Calculate the frequencies of the two ENDOR transitions shown in Fig. 9.2 for a ^{1}H with $a = 10$ MHz recorded on a W-band spectrometer (3.4 T).

Bibliography

General EPR

Atherton, N.M., *Principles of Electron Spin Resonance*, Ellis Horwood Ltd, UK, 1993.

Brustolon, M. and Giamello, E., Eds, *Electron Paramagnetic Resonance: A Practioner's Toolkit*, John Wiley and Sons, New Jersey, 2009.

Eaton, G.R., Eaton, S.S., Barr, D.P., and Weber, R.T., *Quantitative EPR*, Springer Science & Business Media, 2010.

Hagen, W.R., *Biomolecular EPR Spectroscopy*, CRC Press, 2008.

Weil, J.A., Bolton, J.R., and Wertz, J.E., *Electron Paramagnetic Resonance: Elementary Theory and Practical Applications*, John Wiley & Sons, 1992.

Chapter 1

Hore, P.J., *Nuclear Magnetic Resonance*, Oxford Chemistry Primers, 2015.

Rieger, P.H., *Electron Spin Resonance: Analysis and Interpretation*, RSC Publishing, Cambridge, 2007.

Chapter 2

Carrington, A. and McLachlan, A.D., *Introduction to Magnetic Resonance*, Chapman & Hall, John Wiley & Sons, 1979.

Chapter 3

Poole, C.P., *Electron Spin Resonance: A Comprehensive Treatise on Experimental Techniques*, Interscience Publishers, 1967.

Chapter 4

Carrington, A., Electron-spin resonance spectra of aromatic radicals and radical-ions, *Q. Rev. Chem. Soc.*, 1963, **17**, 67–99.

Forbes, M.D.E., Ed., *Carbon Centred Free Radicals and Radical Cations; Structure, Reactivity and Dynamics*, John Wiley & Sons, 2010.

Gerson, F. and Huber, W., *Electron Spin Resonance Spectroscopy of Organic Radicals*, Wiley-VCH, 2003.

Chapter 5

Atkins, P.W. and Friedman, R.S., *Molecular Quantum Mechanics*, Oxford University Press, 2010.

Zare, R., *Angular Momentum; Understanding Spatial Aspects in Chemistry and Physics*, Wiley, 1988.

Pietrzyk, P., Mazur, T., and Sojka, Z., Electron paramagnetic resonance spectroscopy of inorganic materials, in *Local Structural Characterisation*, 1st Edn, Ed. D.W. Bruce, D. O'Hare, R.I. Walton, John Wiley & Sons, 2014.

Chapter 6

Abragam, A. and Bleaney, B., *EPR of Transition Ions*, Dover Publications, 1986.

Atkins, P.W. and Symons, M.C.R., *The Structure of Inorganic Radicals*, Elsevier Publishing Company, 1967.

Mabbs, F.E. and Collison, D., *EPR of d Transition Metal Compounds*, Elsevier, The Netherlands, 1992.

Pilbrow, J.R., *Transition Ion Electron Paramagnetic Resonance*, Clarendon Press, Oxford, 1990.

Chapter 7

Bencini, A. and Gatteschi, D., *Electron Paramagnetic Resonance of exchange coupled systems*, Springer-Verlag, Berlin-Heidelberg, 1990.

Chapter 9

Hoff, A.J., Ed, *Advanced EPR: Applications in Biology and Biochemistry*, Elsevier, 1989.

Hore, P.J., Jones, J.A., and Wimperis, S., *NMR: The Toolkit*, Oxford Chemistry Primers, 2015.

Jeschke, G., DEER distance measurements on proteins, *Annu. Rev. Phys. Chem.*, 2012, **63**, 419–446.

Kevan, L. and Kispert, L.D., *Electron Spin Double Resonance Spectroscopy*, John Wiley & Sons, 1976.

Kurreck, H., Kirste, B., and Lubitz, W., *Electron Nuclear Double Resonance Spectroscopy of Radicals in Solution*, VCH Publishers, 1988.

Schweiger, A. and Jeschke, G., *Principles of Pulse Electron Paramagnetic Resonance*, Oxford University Press, 2001.

Appendices

Appendix A: Units and conversions

Magnetic field

The SI unit of magnetic field induction (or magnetic flux density) is the Tesla (T).

$$1\,\text{Tesla} = 10{,}000\ \text{Gauss}\ (1\,\text{mT} = 10\ \text{G})$$

Hyperfine couplings and splittings

Generally, EPR provides hyperfine splittings (in *field* units). These values are field independent. Hyperfine techniques (ENDOR, ESEEM, HYSCORE) give hyperfine couplings (in *frequency* units). They can be interconverted as follows:

$$A[\text{frequency units}] = \frac{g\mu_B}{h}a[\text{field units}]$$

For organic radicals (when g is very close to g_e), this can be simplified to:

$$A[\text{MHz}] = 28.025a[\text{mT}] \text{ or } A[\text{MHz}] = 2.8025a[\text{G}]$$

For transition metal ions, g can vary significantly, and the following equations should be used:

$$A[\text{MHz}] = 28.025\ (g/g_e)a[\text{mT}] \text{ or } A[\text{MHz}] = 2.8025\ (g/g_e)a[\text{G}]$$

Hyperfine couplings are also quoted in wavenumbers:
A [frequency units] = cA [wavenumbers], where c is the speed of light.
This can be simplified to $A\,[\text{MHz}] = 2.998 \times 10^4 A\,[\text{cm}^{-1}]$ or

$$A[\text{cm}^{-1}] = 9.348 \times 10^{-4}(g/g_e)a[\text{mT}]$$

Anisotropic hyperfine couplings

The anisotropic hyperfine couplings (b_0) quoted in Table 6.1, are reported for an unpaired electron in a p-orbital centred on an atom, and are given by the expression:

$$b_0 = \frac{2}{5}\frac{\mu_0}{4\pi}g_N\mu_N\left\langle r^{-3}\right\rangle_p$$

where the angular brackets integrate over the p-orbitals. Similar adjustments are made for d- and f-orbitals. A word of warning: in the literature, the b_0 values may also be quoted using the coefficient of 4/5 instead of 2/5. One should also bear in mind the sign of g_N when calculating the dipolar hyperfine terms.

Appendix B

Table B.1 Character table for point group D_{4h}.

D_{4h}	E	$2C_4$	C_2	$2C_2'$	$2C_2''$	i	$2S_4$	σ_h	$2\sigma_v$	$2\sigma_d$		
A_{1g}	1	1	1	1	1	1	1	1	1	1		z^2
A_{2g}	1	1	1	−1	−1	1	1	1	−1	−1	R_z	
B_{1g}	1	−1	1	1	−1	1	−1	1	1	−1		x^2-y^2
B_{2g}	1	−1	1	−1	1	1	−1	1	−1	1		xy
E_g	2	0	−2	0	0	2	0	−2	0	0	R_x, R_y	yz, xz
A_{1u}	1	1	1	1	1	−1	−1	−1	−1	−1		
A_{2u}	1	1	1	−1	−1	−1	−1	−1	1	1	z	
B_{1u}	1	−1	1	1	−1	−1	1	−1	−1	1		
B_{1u}	1	−1	1	−1	1	−1	1	−1	1	−1		
E_u	2	0	−2	0	0	−2	0	2	0	0	x, y	

Table B.2 Direct product table for D_{4h}.

	A_1	A_2	B_1	B_2	E
A_1	A_1	A_2	B_1	B_2	E
A_2		A_1	B_2	B_1	E
B_1			A_1	A_2	E
B_2				A_1	E
E					$A_1 + [A_2] + B_1 + B_2$

Appendix C

Hyperfine values for d-block transition metal ions

d^9 case: The predicted A values for the d^9 case (see section 6.2) depend on the ground state in question. For the $d_{x^2-y^2}$ ground state, the $A_{\parallel}$ and $A_{\perp}$ values are given by:

$$A_{zz}\left(A_{\parallel}\right)=P\left[-\kappa+\Delta g_z-\frac{4}{7}+\frac{3}{7}\Delta g_x\right] \quad \text{A.1}$$

$$A_{xx}, A_{yy}\left(A_{\perp}\right)=P\left[-\kappa+\frac{2}{7}+\frac{11}{14}\Delta g_x\right] \quad \text{A.2}$$

In these equations Δg_i is the shift relative to g_e (which as we saw is dependent on symmetry and ground state), P is defined in Chapter 5 (page 55), and κ arises from the Fermi contact interaction associated with finding the unpaired electron at the nucleus. In the case of the d_{z^2} ground state, the $A_{\parallel}$ and $A_{\perp}$ values are given by:

$$A_{zz}\left(A_{\parallel}\right)=P\left[-\kappa+\frac{4}{7}-\frac{1}{7}\Delta g_x\right] \quad \text{A.3}$$

$$A_{xx}, A_{yy}(A_{\perp}) = P\left[-\kappa - \frac{2}{7} + \frac{15}{14}\Delta g_x\right] \qquad \text{A.4}$$

Once again, the key point to note is that the resulting values of A in eqns 6.5–6.8 are different so anisotropic hyperfine patterns will be observed.

d^1 case: The resulting anisotropic A values are given as:

$$A_{zz}(A_{\parallel}) = P\left[-\kappa + \Delta g_z - \frac{4}{7} + \frac{3}{7}(\Delta g_x)\right] \qquad \text{A.5}$$

$$A_{xx}, A_{yy}(A_{\perp}) = P\left[-\kappa + \frac{2}{7} + \frac{11}{14}\Delta g_x\right] \qquad \text{A.6}$$

Index